Marshall Davis Ewell

A Manual of Medical Jurisprudence

For the Use of Students at Law and of Medicine

Marshall Davis Ewell

A Manual of Medical Jurisprudence
For the Use of Students at Law and of Medicine

ISBN/EAN: 9783337311940

Printed in Europe, USA, Canada, Australia, Japan

Cover: Foto ©berggeist007 / pixelio.de

More available books at **www.hansebooks.com**

A MANUAL

OF

MEDICAL JURISPRUDENCE

FOR THE USE OF

STUDENTS AT LAW AND OF MEDICINE.

BY

MARSHALL D. EWELL, M.D., LL.D.

PROFESSOR OF COMMON LAW IN UNION COLLEGE OF LAW, CHICAGO
(LAW DEPARTMENT OF THE NORTH-WESTERN UNIVERSITY).

BOSTON:
LITTLE, BROWN, AND COMPANY.
1887.

UNIVERSITY PRESS:
JOHN WILSON AND SON, CAMBRIDGE.

PREFACE.

THE subject of Medical Jurisprudence is one which
does not receive from schools either of law or of medicine anything like the attention which its importance
deserves. The author, when a student of the law some
twenty years since, and more recently when a student
of medicine, did not have the benefit of a single lec-
ture upon this subject; and it is believed that many, if
not a majority, of the law and medical colleges of the
present day dismiss the subject either with no attention
whatever or with so little attention that students de-
rive no practical benefit from the instruction given.
For a medical man to study the subject (as is usually
the case if he studies it at all) clinically, so to speak,
as the defendant in an action for malpractice or as a
so-called expert witness, while certainly calculated to
make an impression upon his memory, cannot be said to
be wholly agreeable and profitable. On the other hand
it is not uncommon to see attorneys engaged in the
trial of cases involving important interests, betray culpable ignorance of this subject. The attorney, to whom

is intrusted the trial of a case involving an important
question of Medical Jurisprudence, who has a good
knowledge of the leading facts and principles of medicine and of their application to Medical Jurisprudence,
possesses an immense advantage over his adversary who
is not possessed of such knowledge; and no amount of
preliminary "cramming" will supply the place of previous general knowledge of the whole subject. It is
the deliberate judgment of the author that no man can
be said to be prepared for every emergency which may
arise in the practice of either law or medicine without
a very considerable acquaintance with the science of
Medical Jurisprudence.

There are so many uniformly excellent treatises on
this subject that there seems to be no need for another;
but these works are, as a rule, so large as to take more
time for their perusal than the student at law or of
medicine has to bestow upon them; moreover, most
of them are quite expensive. The experience of the
writer, both as a student and teacher, has led him
to believe that a work which within a moderate compass states all the leading facts and principles of the
science, concisely and yet clearly, will prove useful to
students of both professions; this work was accordingly
undertaken. It does not claim to be more than a careful compilation from what seemed to be the best authorities of the leading principles of the science. It

is believed, however, that within the moderate compass of this volume will be found the substance of all the principles stated in the more voluminous works. Two exceptions should be made to this statement as respects the subjects of Insanity and Toxicology. To treat either of these subjects with anything like completeness would require more space than is occupied by the whole volume; it was therefore thought better by the author to treat with a reasonable degree of completeness the other subjects usually discussed in works upon Medical Jurisprudence, giving an outline only of these two, than by attempting to cover the whole ground, including Toxicology and Insanity, to make the whole meagre and unsatisfactory. Accordingly, upon the subject of Insanity a brief outline only has been given, which, however, it is hoped, will be found to accord with the best authorities of the day. As respects Toxicology, nothing more has been attempted than a consideration of some general principles leading up to the more particular study of the subject. The reasons for this will be found more fully stated at the beginning of the chapter upon that subject.

In order to save space it has been thought expedient to omit the citation of authorities for the different statements of the text, and, as a rule, they have therefore been omitted. It is proper, however, in this place to say that free use has been made of the standard

European works upon this subject, and that the author
is especially indebted to the valuable works of Ogston,
Tidy, Taylor, and Woodman & Tidy; special acknowl-
edgments are also due to Dr. Spitzka, of New York,
of whose manual upon Insanity free use has, with the
author's consent, been made.

MARSHALL D. EWELL.

UNION COLLEGE OF LAW, CHICAGO,
    Aug. 2, 1887.

CONTENTS.

———◆———

MEDICAL JURISPRUDENCE.

CHAPTER I.

INTRODUCTORY; EVIDENCE; EXPERTS; COMPENSATION;
RELATION OF PHYSICIAN TO PATIENTS, ETC.

WHILE the term "medical jurisprudence" is a misnomer, — the collection of facts and conclusions usually
passing by that name being principally only matters
of evidence, and rarely rules of law, — still, the term is
so generally employed that it would be idle to attempt
to bring into use a new term, and we shall accordingly
continue the employment of that which has only the
sanction of usage to recommend it.

There are, however, **some rules of law** pertaining to
medical witnesses, and to a few other topics, which
may properly be treated in a work like the present.
No space, however, will be occupied in discussions,
often to be found in works upon medical jurisprudence,
of rules of law which have no especial reference to
medical practitioners more than to other citizens.

The jurisdiction and practice of the various courts,
and the rights and duties of witnesses vary to some
extent in the different States, and will be found defined
in local works of practice and the statutes of the differ-

ent States, as well as to some extent in general treatises
upon the common law, and it is beyond the scope of
this work to consider them.

It may not be amiss, however, to state that the
subpœna of a court of justice cannot be disregarded
with impunity, but should be obeyed promptly. A
failure to obey may subject the witness to attachment
for contempt.

The medical witness should remember, also, that by
the common law **a medical man has no privilege to
avoid giving in evidence** [1] any statement made to him
by a patient; but when called upon to do so in a court
of justice he is bound to disclose every communication,
however private and confidential, which has been made
to him by a patient while attending him in a professional capacity. By statute, however, in some of the
United States, communications made by a patient to
a physician, when necessary to the treatment of a case,
are privileged, and the physician is either expressly
forbidden, or not obliged to reveal them. Such statutes
exist in Arkansas, California, Indiana, Iowa, Michigan,
Minnesota, Missouri, Montana, New York, Ohio, and
Wisconsin. The seal upon the physician's lips is not
taken away by the patient's death. Such communications, however, must be of a lawful character and not
against morality or public policy; hence, a consultation
as to the means of procuring an abortion on another is
not privileged, nor would be any similar conference
held for the purpose of devising a crime or evading its
consequences.

[1] See *Jones* vs. *King*, cited *post*, in this chapter.

A report of a medical official of an insurance company, on the health of a party proposing to insure his life, is not privileged from production; nor is the report of the surgeon of a railroad company as to the injuries sustained by a passenger in an accident, unless such report has been obtained with a view to impending litigation.

Compensation. — By the Roman law the services of both an advocate and of a physician were strictly honorary; by the common law of England, surgeons and apothecaries could recover by law remuneration for their services, but a physician was presumed to attend his patient for an honorarium, and could not sue therefor until the passing of the medical act in 1858.

By the law in this country all branches of the profession may recover at law a reasonable compensation for their services, the amount of which, unless settled by law, is a question for the jury; in settling which, the eminence of the practitioner, the delicacy and difficulty of the operation or of the case, as well as the time and care expended, are to be considered. There is no limitation by the common law as to the amount of such fees provided the charges are reasonable. The existence of an epidemic does not, however, authorize the charge of an exorbitant fee.

A medical man can also recover for the services rendered by his assistants or students, even though the assistant is unregistered; it is not necessary that there should be any agreed specified price, but he will be allowed what is usual and reasonable.

It is not the part of a physician's business, ordinarily, to supply the patient with drugs; if he does so he has a right to compensation therefor. If the agreement is, "no cure, no pay," he cannot, however, even recover for medicines supplied, if the cure is not effected.[1] His right to recover for professional services does not depend upon his effecting a cure, or upon his services being successful, unless there is a special agreement to that effect; but it does depend on the skill, diligence, and attention bestowed. The practitioner must be prepared to show that his work was properly done, if that be disputed, in order to prove that he is entitled to his compensation. See *post*, MALPRACTICE. Where the surgical instruments employed in amputating an arm were a large butcher-knife, and a carpenter's sash-saw, a charge to the jury that if the operation was of service and the patient did well and recovered, the surgeon was entitled · to compensation, — although it was not performed with the highest degree of skill, or might have been performed more skilfully by others, — was held proper.

If a surgeon has performed an operation which might have been useful, but has merely failed in the event, he is nevertheless entitled to compensation; but if it could not have been useful in any event, he has no claim upon his patient. A medical man who has made his patient undergo a course of treatment which plainly could be of no service cannot make it a subject of charge; but an apothecary who has simply administered medicines under the direction of a physician may re-

---

[1] See the late case of *Jones* vs. *King*, S. Ct. Ala., 1 South. Rep. 591 ; 24 Cent. L. J. 434. In this case it was held competent to prove that the plaintiff's medicines were worthless, and, moreover, that not being patented he had no property in the secret of his remedy such as the law would privilege him from disclosing.

cover for the same, however improper they may have been.

The number of visits required must depend upon the circumstances in each particular case, and the physician is regarded by the law as the best and proper judge of the necessity of frequent visits; and in the absence of proof to the contrary it will be presumed that all professional visits made were deemed necessary and were properly made.

There must not be too many consultations. The physician called in for consultation or to perform an operation may recover his fees from the patient, notwithstanding that the attending practitioner summoned him for his own benefit, and had arranged with the patient that he himself would pay.

Where a medical man has attended as a friend he cannot charge for his visits.

Where a tariff of fees has been prepared and agreed to by the physicians of any locality, they are bound by it legally as far as the public are concerned (that is to say, they cannot charge more than the tariff rates), and morally as far as they themselves are concerned.

A physician is always allowed discretionary powers over his patients as to his mode of treatment, so as to be able to alter the same according to the varying necessities of the case. Unless such change of treatment involves a risk of life, or consequences of which he is unwilling to assume the responsibility, he is not under obligations to give notice or obtain permission before making it.

It is the duty of a physician who is attending a patient having a contagious disease, when called upon to attend others, to take all such precautionary means as experience has proved to be necessary to prevent its communication to them. Where a physician who was told by a patient not to attend any person infected with small-pox, or his services would be dispensed with, failed to say that he was attending such a patient and promised not to do so, but continued so to attend, and by want of proper care communicated small-pox to plaintiff and his family, — these facts were held proper evidence to go to the jury in reduction of damages in an action for his account, and the physician was held responsible for the suffering, loss of time, and damages to which the patient was subjected. If a physician by communicating an infectious disease has rendered a prolonged attendance necessary, thereby increasing his bill, he cannot recover for such additional services necessitated by his own want of care.

As to who should pay the physician, but little can be stated in this connection; a few rules and principles however, are the following: If A says to B, a medical man, "Attend upon C, and if he does not pay you, I will," this being a promise to answer for the debt of C, for which he is also liable, the promisor cannot according to some authorities be held unless the promise is in writing, signed by A or by some one thereunto by him lawfully authorized. But if he says to B, "Attend C, and charge your bill to me;" or, "I will pay you for your attendance upon C," — no written agreement is necessary.

A person who calls upon a physician and directs him to attend upon a patient may render himself liable for the fees of the physician; the question in such a case is whether the party so calling the physician is or is not a mere agent; if he is a mere agent, he is not liable, but otherwise he is ordinarily liable.

A wife has implied authority to bind her husband for reasonable expenses incurred in obtaining medicines and medical attendance; but this implied authority is terminated if she leaves home of her own accord without sufficient reason, and the fact has become notorious, or the husband has given sufficient notice that he will no longer be responsible for debts of her contracting. If, however, a husband turns an innocent wife out of doors without the means to obtain necessaries, he sends his credit with her, and in such case medical attendance is undoubtedly a necessary. If a physician attends a wife whom he knows to be living separate and apart from her husband, it is his duty to inquire whether she has good cause for so doing; if she has not, he cannot collect his bill from the husband.

Although the law favors no particular school of medicine, it does not encourage mere quackery. Where, therefore, a so-called doctor was in the habit of putting a woman into a mesmeric sleep, who thereupon became clairvoyant and prescribed medicines, which the doctor furnished and for which he sued, the judge, in deciding the case said : —

" The law does not recognize the dreams, visions, or revelations of a woman in mesmeric sleep as necessaries for a wife, for which the husband without his consent can be made to pay."

By the common law the duty of a father to furnish necessaries, including medical attendance, for his minor child, is a moral obligation and not a legal one. The father's liability, if it exists, is put on the ground of agency, and the authority of an infant to bind the father by contracts for necessaries, including medical aid, will be inferred from very slight evidence. Medicines and medical aid are necessaries, for which an infant may, when not otherwise supplied, legally contract, and for which he may render himself liable.

A master is not bound to provide medical assistance for his servant; the obligation, if it exists at all, must arise from contract, and such contract will not be implied from the fact that the servant lives under the master's roof, nor because the illness of the servant has arisen from an accident met with in the master's service.

A master is bound to provide an apprentice with proper medicines and medical attendance.

Where a physician or surgeon has been called in to attend a passenger or employee injured by a collision or other railway accident, the company cannot be held responsible unless it can be shown that the agent or servant who summoned the medical man had authority so to do. It has been held that neither a guard nor the superintendent of a station nor the engineer of the train in which the accident happened has any implied authority to bind the company for such medical services. In England it has been held that the general manager of the railway company has, as incidental to his employment, authority to bind his company for medical services bestowed upon one injured on his railway. In Illinois a similar decision has been rendered as to a general superintendent, but a contrary decision has been rendered by the Superior Court of the city of New York.

The Compensation of Medical and other Experts is a subject which has been somewhat discussed by the courts, and upon which the law is not entirely settled.

Statutory provisions will be found in some of the States regulating this question. Thus in Iowa it is enacted that —

" Witnesses called to testify only to an opinion founded on special study or experience in any branch of science, or to make scientific or professional examinations, and state the results thereof, shall receive additional compensation, to be fixed by the court, with reference to the value of the time employed and the degree of learning or skill required."

Similar provisions, more or less extensive, are to be found upon the statute-books of North Carolina, Rhode Island, and Minnesota.

In Indiana, by statute, experts are compellable to testify to an opinion without extra compensation.

While it is the general practice of parties employing experts to give them extra compensation, it is regarded as having been paid for the party's own benefit, and hence can not, in the absence of statute, be regarded as a necessary disbursement, and hence taxed as costs. In every State there can, we think, be no doubt that an expert cannot be compelled to make a preliminary examination, such as the analysis of a stomach or the examination of an alleged lunatic, so as to enable the expert to testify in court as to his professional opinion, without special compensation. It would seem also that an expert cannot be required to attend during an entire trial, for the purpose of listening to the testimony, with the view of enabling him to express an opinion thereupon, without special compensation.

In the absence of statute and of such preliminary labor, can an ordinary expert witness legally require the payment of special compensation as a condition precedent to his testifying? When testifying as ordinary witnesses to facts which have fallen under their notice, they stand upon the same basis as ordinary witnesses. When testifying as to matters requiring professional skill, the question is not so clear. In England, at least in civil cases, additional compensation seems to be necessary. But in England, a professional man, even when called as to facts and not opinions, is entitled to extra compensation on the higher scale allowed under the statute of Elizabeth (5 Eliz. c. 9), which provides that the witness must " have tendered to him, according to his countenance or calling, his reasonable charges."

Writers upon medical jurisprudence have generally been of the opinion that witnesses are entitled to extra compensation as a matter of right; but these authors have been, almost without exception, medical men, and their opinion, except so far as supported by adjudicated cases, is not conclusive in deciding a question like the present. Passing by, therefore, -these writers, a study of the authorities seems to establish the following propositions : —

In the first place, as already stated, unquestionably medical men are not, in this country, when testifying as ordinary witnesses to facts within their own personal knowledge, entitled to extra compensation, even though their professional skill may have enabled them to observe such facts more intelligently; and this is so even in those States allowing extra compensation to expert witnesses when called as experts.

The general practice is thus stated by Mr. Greenleaf in his work on Evidence : —

"In order to secure the attendance of a witness in civil cases, it is requisite by statute, 5 Eliz. c. 9, that he have tendered to him according to his countenance or calling, his reasonable charges."

Under this statute it is held necessary in England that his reasonable expenses for going to and returning from the trial, and for his reasonable stay at the place, be tendered to him at the time of serving the subpœna ; and if he appears he is not bound to give evidence until such charges are actually paid or tendered, unless he resides and is summoned to testify within the weekly bills of mortality, in which case it is usual to leave a shilling with him upon the delivery of the subpœna ticket. These expenses of a witness are allowed pursuant to a scale graduated according to his situation in life. But in this country these reasonable expenses are settled by statutes, at a fixed sum for each day's actual attendance, and for each mile's travel from the residence of the witness to the place of trial and back, without regard to the employment of the witness or his rank in life. The sums are not alike in all the States, but the principle is believed to be everywhere the same.

"In some States it is sufficient to tender to the witness his fees for travel from his home to the place of trial, and one day's attendance in order to compel him to appear upon the summons; but in others the tender must include his fees for travel in returning.   Neither is the practice uniform in this country as to the question whether the witness, having appeared, is bound to attend from day to day until the trial is closed, without the payment of his daily fees ; but the better opinion seems to be that without payment of his fees, he is not bound to submit to an examination.

"In criminal cases no tender of fees is, in general, necessary on the part of the government in order to compel its witnesses to attend, — it being the duty of every citizen to obey a call of this description, and it being also a case in which he is himself in some sense a party.   But his fees will, in general, be finally paid from the public treasury.   In all such cases the accused is entitled to have compulsory process for obtaining witnesses in his favor."

Ex parte Dement, 53 Ala. 389, decided in 1875, is a well-considered case upon the question under consideration.   In this case it was held that a physician, like any other person, may be called to testify as an expert in a judicial investigation, whether it be civil or criminal in its nature, without being paid for his testimony as for a professional opinion ; and that upon refusal to testify he may be punished as for a contempt.   This case will repay a careful perusal, as in it the authorities are well collected and discussed.

The principle of the case of *Ex parte* Dement was approved and followed in the late case of *The State* vs. *Teipner* (S. Ct. Minn.), 32 N. W. Rep. 678, decided May 6, 1887.

In *Summers* vs. *The State*, 5 Tex. App. 365 (1879), it was likewise held that, although a physician cannot be compelled to make an autopsy, yet, having made it, he may be compelled to testify as to its results.

In Illinois the rule seems to be settled by the case of *Wright* vs. *The People*, 112 Ill. 540 (1884). In this case the physician having voluntarily stated his profession, etc., and the symptoms of the plaintiff in a civil action for assault and battery, refused to answer the following question calling for a professional opinion, unless his fee of $10 should be paid or secured to him:—

"*Question:* If one person should strike another a heavy blow on the head, at or near the temple, with that billy, would it or would it not be likely to produce upon the person receiving such blow a condition alike or similar to that in which you find Jno. Finneran."

Upon the witness's refusing to answer, the court below fined him as for a contempt, and the Supreme Court affirmed the judgment. The court, however, in their opinion do not go beyond the case stated, and perhaps may rule differently upon a different case.

Opposed to these cases are *Buchman* vs. *The State*, 59 Ind. 1; *Dills* vs. *The State*, id. 15 (following *Blythe* vs. *State*, 4 Ind. 525). In these cases the court, upon general principles, came to a conclusion opposite to that of *ex parte* Dement, but also place their decision upon Sec. 21 of the Bill of Rights of Indiana, which provides that "No man's particular services shall be demanded without compensation." The authority of *Buchman* vs. *State*, and *Dills* vs. *State*, is weakened by the fact that two, Biddle, C. J., and Niblock, J., of the five judges dissented.

In re Roelker, 1 Sprague, 276 (1855), (a *nisi prius* case) holds that the court will not compel the attendance of an interpreter or expert who has neglected to obey a subpœna, unless in a case of necessity, and does not seem to be a direct authority upon this question.

In *U. S.* vs. *Howe* (U. S. Dist. Ct., Western Dist. Ark., 12 Cent. Law Jour. 193) the court refused to punish as for a contempt a Dr. Bennett, who refused to testify unless first paid a reasonable compensation.

The Hon. Emory Washburn, in discussing this subject in an address before the American Academy of Arts and Sciences (1 Am. Law Rev., 1866, p. 63), used the following language : —

"If the case be one of a public nature, involving the question of a crime of magnitude, where the public safety requires the investigation, the right to compel the attendance of such witnesses becomes an incident to the exercise of government itself, in the same way that a juror is obliged to sacrifice convenience or profit to render a public service, or the soldier is called upon to take up arms in defence or execution of the law. It rests upon the maxim *salus populi suprema lex.*"

He then quotes approvingly the following from 1 Greenl. Ev. § 310, note : —

"There is also a distinction between a witness to facts and a witness selected by a party to give his opinion on a subject with which he is peculiarly conversant from his employment in life. The former is bound as a matter of public duty to testify to facts within his knowledge. The latter is under no such obligation; and the party who selects him must pay him for his time before he will be compelled to testify." *Webb* vs. *Page*, 1 C. & K. 23.

It will be observed upon careful examination that nearly all the authorities holding that a physician cannot be compelled to testify as an expert without extra compensation proceed upon the authority of the case of *Webb* vs. *Page, supra,* or are influenced by statutory or constitutional provisions.

According to the weight of authority it seems that in those States where there is no statute or decision settling the question in favor of the physician, his only prudent course is to testify if ordered by the court to do so, without demanding extra compensation, as otherwise he runs the risk of punishment for contempt.

As to special contracts for compensation, they are subject to the limitation that the contract cannot be made conditional upon the success of the suit in which the expert is to testify. A contract thus conditioned is against public policy and void.

It is also unlawful for a physician to make a contract under which he examines into the condition of a person injured by a railway or other accident and reports the same to the party liable, under a stipulation that his compensation shall be proportioned to the amount recovered from the defendant. Such a contract is void as being against public policy.

As respects the Relation existing between Physicians or Surgeons and their Patients, the rule is well settled that where one occupies a position which naturally gives him the confidence of another or which in any way gives him an influence or an undue advantage over the other, transactions between them require something more to give them validity than is necessary in other cases. In such case where a physician or surgeon has obtained from his patient any sort of valuable security or property, he takes upon himself, if the validity of the

transfer is attacked, the burden of showing that no undue influence was exercised by him over the patient,
and of establishing the perfect fairness and equity of the
transaction; if he cannot establish these facts the transaction will not be allowed to stand.

In the case of *DeMay* vs. *Roberts*, 46 Mich. 161, a
physician took an unprofessional friend with him to
attend a case of confinement when there was no emergency requiring the latter's presence. The physician
told the patient's husband that he had brought a friend
along with him to help him carry his things, and he
was accordingly admitted. The patient, on afterwards
discovering the fact, sued both, and it was held that the
plaintiff and her husband had a right to presume that
the outsider was a medical associate; that in obtaining
admission without disclosing his true character, the
defendants were guilty of deceit; that plaintiff had a
right to testify that she supposed he was a physician or
medical assistant, and also to give evidence of whatever
may have been said at the time tending to support such
supposition. It was also held in the same case admissible to ask a competent witness as to the custom
among physicians in regard to calling assistance in
these cases.

CHAPTER II.

The Importance of a Proper Order and Method as well as thoroughness in making a medico-legal inspection of a dead body and its surroundings cannot well be over-estimated; as, if done in an unsystematic or incomplete manner, it will not unfrequently defeat rather than aid in the administration of justice. To use a technical expression, the degree of certainty as to the cause of death should, when possible, be "certainty to a certain intent in every particular;" that is to say, the examination should be so complete and thorough as not only to assign an adequate cause of death, but to negative any other cause. While this degree of certainty is perhaps not attainable in many instances, it is in some; and it should be the constant endeavor of the medico-legal practitioner to approach as nearly as possible to it; and coroners and other officers engaged in the preliminary examinations necessary for the detection and punishment of crime should tolerate nothing less.

Medico-Legal Inspections which are expected to come before the courts should be made by at least two properly qualified medical men, and within from twenty-four to forty-eight hours after death, whenever practicable. When necessary, however, to advance the interests of justice, they may be made at any time after death; even putrefaction will not afford any valid reason for not undertaking the same. They should, if possible, be made by natural light, as when made by artificial light certain characteristic colors indicative of poisoning would probably escape notice. The body, if frozen, should be allowed to thaw before commencing the examination, by leaving it for some time in a warm room.

The surroundings of the body should first be carefully noted, and if possible a plan made locating the position of the body when first seen, with reference to other surrounding objects, such as furniture, instruments of violence, cups, bottles, etc. Any blood-stains on the person, clothing, or elsewhere, should be carefully noted; and articles likely to be important as means of evidence should be marked and carefully preserved so as to be capable of future identification.

The body should be first examined before the clothes are removed; the state of the clothes, whether marked with blood-stains, etc., cut, or torn, — and if cut, the exact position of the cuts and their relation to the wounds on the body, if any, — should be carefully noted, and the clothes themselves preserved for use as evidence.

Next, the attitude and position of the limbs, the condition of the hands and nails as affording evidence of a struggle before death, should be noted; also the color and other characteristics of the hair on the head and face; the condition of the teeth, mouth, and tongue; color of the eyes and pupils; and the expression and color of the face. Particular notice should be taken whether or not the body is in a condition of post-mortem rigidity.

Having completed the external examination, the clothes should be removed; the body should then be examined as to its temperature, the presence of rigidity, its state of putrefaction or otherwise, the sex, height, and weight, probable age, development, whether lean or fat, and the general condition of the body as to the state of its nourishment. The body should also be examined with reference to the color of the skin, scars, and marks thereon, abnormities of structure, and stains, such as feces, semen, etc. Careful examination should be made for injuries, such as contusions, wounds, etc.; if any wounds are found, their nature, depth, direction, and extent should be carefully determined. The condition of the edges of the wounds should be particularly examined, and their depth carefully determined, not by probing, but by careful dissection. The neck should be examined to see if there are any marks of external violence indicative of strangulation, etc.; if a cord is found its position upon the neck or otherwise should be carefully noted.

In examining the body for injuries, particular examination should be made of all the inlets and outlets of the body. It should be remembered that a fatal wound may be inflicted through the orbit, fontanelles in the case of infants, the mouth, vagina, or anus.

If the subject to be examined is an infant, its length and weight, the length and state of the hair, length and condition of the nails, presence or absence of the membrana pupillaris, condition of the genital organs, the centres of ossification, the umbilical cord, the fontanelles, and cartilages of the nose and ears, should all be carefully noted.

In making a post-mortem examination a knowledge of pathological anatomy is of the greatest importance. As has been well observed by Virchow, —

"Medico-legal technics, with all due deference to the independence of forensic medicine, will always go hand in hand with pathological anatomy, for the latter is more universal: it has to deal with cases of all kinds, and for that reason is a great protection against that one-sidedness with which medico-legal practice is so much encumbered."

A well considered **order of examination** is here of the greatest importance; and an examination of the various authors upon this subject reveals a considerable variety of methods. Many authors advise the examination of the head first. Casper recommends opening first that cavity in which there is the greatest probability of finding the cause of death. While exceptional cases may necessitate a change of this order, the following scheme followed by Professor Virchow has much to recommend it. His practice has for years been that "under all circumstances the abdomen is to be first opened, but not dissected." It is only necessary in this stage to determine the position of the diaphragm and other organs; and to note any abnormal contents of the abdomen which may possibly be present, and the color of the parts exposed. The position of the abdominal viscera and their relations should be determined by manual examination; their color should be noted, remembering that the color of many parts is entirely due to the blood and that in a dead body the arterial blood cannot be well distinguished from venous blood. The presence or absence of pathological coloring-matters, such as pus or bile, and of extravasations should be noted. If any foreign matters are found, careful note of the same should be made. The presence of adhesions and other pathological conditions which may be determined without dissection, should also be noted. The position of the diaphragm in the dead bodies of infants should be noted immediately after the opening of the abdomen.

After the completion of this preliminary examination of the abdomen, the thorax is immediately to be opened and dissected unless there is some cogent reason for departing from the rule. The suspicion of poisoning is always admitted to be a reason of this nature, and in this case the whole of the examination centres in the stomach, and every precaution must be taken to place it and its contents without loss or change at the disposal of the law. See *post*, TOXICOLOGY.

In examining the thorax the condition of the pleural cavities should be noted, as respects the position, color, and condition of their contents, and presence of foreign bodies; and this without cutting any of the important structures, such as arteries, veins, etc.

The pericardium and heart should next be examined. As observed by Professor Virchow, —

"He who would open the pericardium and dissect the heart before determinining whether hæmato-thorax, hydrothorax or pleuritis be present, is a man who ought not to undertake a preliminary examination at all."

The lungs ought not to be removed from the thorax before the heart has been examined, for this cannot be done without separating the pulmonary arteries and veins. Before removing any of the viscera from the thorax for more detailed examination, ligatures should be placed upon all the important vessels connecting the same with the cranial and abdominal cavities, and the sections of the vessels made between two ligatures.

In the examination of the contents of the thorax, note any adhesions of the lungs, — whether they fill the chest as in emphysema, or are collapsed, — the presence or absence of tubercular or other disease, etc. In the examination of the pericardium, note the presence, nature as purulent, sero-purulent, or serous, of the contents of the sac. Note also the presence of tumors, if any, in the thorax.

In examining the heart and vessels note the size of the heart, fulness of the coronary vessels and of the different cavities of the heart, weight of the heart, its condition as to fatty degeneration, etc., hypertrophy, condition of the valves, etc.; the condition of the vessels whether atheromatous or having aneurisms. As to the lungs, note the color, nature of the surface, capacity for air, character of the lung tissue, and of the fluid they exude on pressure; condition of the bronchial tubes and pulmonary artery; the presence of foreign matter in the air passages, and any pathological conditions not above enumerated. The method of examining the lungs of new-born children will be considered in the chapter on INFANTICIDE.

Prolonging the incisions to the chin, the larynx, trachea, pharynx, œsophagus, and the great vessels of the throat should next be examined.

Returning now to the further inspection of the abdominal organs, they should, according to Professor Virchow, be examined in the following order: —

1. The Omentum. 2. The Spleen. 3. The Left Kidney, Supra-Renal Capsule, and Ureter. 4. The Right Kidney, Supra-Renal Capsule, and Ureter. 5. The Bladder, Prostate Gland, Vesiculæ Seminales, and Urethra. 6. (a) The Testicles, Spermatic Cord, and Penis; (b) The Vagina, Uterus, Fallopian Tubes, Ovaries, and Parametria. 7. The Rectum. 8. The Duodenum, Portio Intestinalis of the Ductus Communis Choledochus. 9. The Stomach. 10. The Hepato Duodenal Ligament, Gall-Ducts, Vena Portæ, Gall-Bladder, and Liver. 11. The Pancreas, Cœliac (Semi-Lunar) Ganglia. 12. The Mesentery with its glands, vessels, etc. 13. The Small and Large Intestine. 14. The Retro-Peritoneal Lymphatic Glands, Receptaculum Chyli, Aorta, Vena Cava Inferior.

After having completed the examination of the thoracic and abdominal viscera, the brain and the spinal cord should next be examined. In examining the brain, it may be necessary in some cases to shave the whole or a portion of the head. Having done so when necessary, an incision should be made across the head from ear to ear, and the scalp reflected; any injuries to the scalp should be noticed, and the external surface of the skull should be carefully examined for fractures or other injuries. The skull should then be carefully sawn around, about one-half inch above the meatus auditorius externus, and the calvaria carefully removed. The condition of the dura-mater, arachnoid, pia-mater, the great longitudinal sinus, and the surface of the cerebral hemispheres so far as exposed, should be carefully examined and described. If the dura-mater is adherent to the skull-cap it may be divided before detaching the latter, and the skull-cap removed with the dura-mater adher-

ing to it, as otherwise in endeavoring to separate the
skull-cap from the adherent dura-mater there is danger
of tearing the latter and crushing the brain itself.   In
new-born infants and children these parts are, as a rule,
adherent, rendering necessary the procedure above described.   The brain should next be carefully removed,
and sliced from above downwards, and examined with
particular reference to congestions, extravasation, effusions, serum, blood, lymph, pus, aneurism, embolism,
tumors, etc.   These slices should be *thin*, as within the
interior of a section five millimetres thick there is
ample room for foci of morbid material sufficient to produce paralysis or convulsions.   The less is found, the
greater should be the number of sections.   The condition
of the internal portion of the skull, as to fractures, etc.,
should also be carefully examined, as well as any mechanical injury to the brain corresponding thereto.

When for any reason the brain is first examined, the
head should be raised upon a block as high as can be
done without stretching the parts about the nucha,
and the head then opened in the usual way, and the
brain examined *in situ*.   The head should be allowed
to remain in the same position till the condition of the
abdominal and thoracic organs is ascertained; or if lowered, an assistant should compress the vessels at the base
of the skull to prevent any draining of blood from the
carotid or vertebral arteries or jugular veins.   For this
reason it will usually be expedient to examine the abdominal and thoracic organs before the head, which
with the spinal cord may be last examined.

To examine the spinal cord and its membranes, the
vertebral laminae should be carefully sawn through,

and the cord removed with its dura-mater uninjured; the external and internal condition of the membranes should be noted, and sections made throughout the entire length of the cord, and all injuries and pathological conditions noted. For further details as to the more particular examination of the separate organs, see Virchow's "Post Mortem Examinations" and the larger treatises upon medical jurisprudence.

In making all of the above described examination, carefully written notes should be made as the work progresses; particular cases · may render necessary a deviation from the above-described order of examination, but it should be adhered to in every case where there is no sufficient reason to the contrary.

CHAPTER III.

The definition of a wound, both in surgery and at law, is involved in considerable confusion. In surgery it is usually considered to mean a solution of continuity in any part of the body, suddenly made by anything that cuts or tears, with a division of the skin; by which some understand is meant not only the external cutis, but also the inward membranes of the gullet, intestines, bladder, urethra, and womb. The legal definition of the term "wound" is usually considered to be a breach of the skin, or of the skin and flesh, produced by external violence. But the meaning of the term is often modified by the context and the subject to which it is applied; and it will be convenient in this connection to include within the term any personal injury suddenly arising from any kind of violence applied externally, whether such injury is external or internal. It will also be convenient in this connection to consider the subject of "Burns and Scalds."

Wounds are classified as Punctured Wounds, Incised Wounds, Contused Wounds, Lacerated Wounds, and Poisoned Wounds.

A punctured wound is one such as is usually caused by a pointed instrument, such as a knife, sword, or bayonet; and its depth will therefore be much greater than its superficial extent.

An incised wound, commonly known as a cut, may be defined as a solution of continuity without loss of substance, — such as is usually produced by some cutting instrument with a more or less sharp or thin edge.

A contused wound is one in which there is more or less bruising, and hence discoloration of the skin or tissues from the effusion of blood from small ruptured vessels into the surrounding tissue.

A lacerated wound is one which may be produced by a blunt or dull instrument, stones, bullets, missiles from firearms, etc. Many wounds are of a mixed character, such as those made by firearms or by the use of a knife, in which at times one part of the wound may be incised and another part lacerated.

Poisoned wounds may possess either of the above characteristics, with the addition of some poisonous matter introduced therein at the time the injury is inflicted; they are usually made by cutting or pointed instruments, or by the stings or bites of insects, reptiles, etc.

Punctured wounds. In punctured or penetrating wounds as above defined, the margins will generally not be found in close contact, but somewhat apart, and the regularity of their edges will vary with the sharpness of the instrument producing them; when following thrusts with instruments properly so called, they may usually be distinguished from those caused by glass, crockery, nails, etc., by the sharpness of their edges, their freedom from contusions, and their amount of retraction or the reverse; however occasioned, they usually bleed but sparely, — unless when some large vessels are divided or where cavities are .penetrated, — and almost never heal by first intention. They may be produced by any weapon or other substance having a sharp point, such as a thin cane, tobacco-pipe, arrows, sharp stones, etc. The solutions of continuity do not always take the shape of the instrument by which they were produced, but vary not only with the nature of the instrument, but with the situation or direction of the wound, character of the tissue, its degree of tension, etc. The wound is seldom as large as the instrument which caused it. The openings made in the clothes before entering the body are in this class of wounds generally smaller than the instrument producing them. Punctured wounds, even when made with a sharp instrument, are not always in the form of a straight cut, but not infrequently send off a spur somewhere in their course like the two lines of the Greek letter γ.

Incised wounds. Although these wounds are usually made by instruments with cutting edges more or less sharp, they are not infrequently made by articles having an angle nearly equal to or even exceeding 90 degrees; even the fist, or a blunt body striking a sharp bony ridge, will often produce sharply cut, gaping wounds. The superficial extent of these wounds is usually greater than their depth, and they usually bleed more freely than all others. Their margins are generally sharp, straight, and well defined. It is to be remarked, however, that even where a sharp cutting instrument is used, if considerable force has been employed it will sometimes be found not only to have cut, but also to have bruised and lacerated the parts divided. Incised wounds gape more than other wounds, but the state of the muscles as to contraction or relaxation has considerable influence upon the size of the wound. An incised flesh-wound in the direction of the fibres of the muscle will give rise to little or no gaping of the wound, while a transverse cut will appear deeper and will gape more. It is not always possible to determine whether an incised wound was produced by a more or less blunt body or a cutting instrument, especially where from twenty-four to forty-eight hours have elapsed, and the wound has not healed by first intention. In attempting to determine the kind of instrument employed, the condition of the edges should be particularly noticed; which, however, can only well be done while the wound is recent. Where the wound is produced by a sharp instrument drawn across the part, the edges will be found to be straight; but if the instrument is blunt, or pressure has been used upon it against the part, the edge will be found more or less serrated or irregular. In incised wounds where the different parts are nearly of uniform consistency, the deepest part of the wound is commonly nearer the commencement than the end. At the commencement of incised wounds, not infrequently one or two slight superficial incisions have been made before the person was able to inflict the principal wound. The principal wound frequently ends in a bifurcation or in several points. These distinctive characteristics are usually more noticeable in wounds of the throat; though even here they are not invariable in their occurrence and are most apt to be wanting in cases of suicidal cut-throat with blunt instruments.

In the case of homicidal or suicidal wounds in the living, the principal incision is rarely the only one, and the line of the incision will diverge more or less from the straight course which usually characterizes an incision in a dead body.

Where incised wounds are made through the clothing, the cuts through the same are usually larger than the wounds.

As to the appearance assumed by incised wounds not healing by first intention, it may be laid down as a general rule that during the first twelve hours after their infliction such wounds will be bloody. About this time inflammation will have commenced, with secretion of serosity, which will continue during the second day. The third day sero-purulent matter will begin to exude, and by the fourth or fifth day or even later, suppuration will be fully established, which in a simple wound without loss of substance may last from five to eight days; and from the fifteenth to the eighteenth day the wound will cicatrize. This is the rule laid down by Dr. Ogston; but when the wound is treated according to the modern system of aseptic surgery this rule will require more or less modification. The time required for healing will be considerably lessened, and there may be (when the case is properly treated) no suppuration.

The cure of an incised wound will, however, under ordinary treatment, be influenced very much by circumstances. Thus, it may heal by first intention in some healthy persons, without any suppuration; and where the subject is unhealthy a much longer time may be taken in effecting a cure. The result may also be influenced by the age of the party, depth and locality of the wound, and other circumstances.

An important question frequently arises in this connection, whether an incised wound has or has not been inflicted during life. The experiments of Orfila and of Taylor show that the appearances of wounds inflicted immediately after death so closely resemble those of wounds inflicted immediately before death as not to be distinguishable from each other; but when the infliction of the respective wounds dates either a few hours before or after death, there are some marked appearances by which the time of their infliction may usually be determined. Incised wounds made by sharp instruments on living persons usually bleed more or less freely, and the blood will be found to have clotted on various parts of the body, clothes, floor, or surrounding objects. If the wound has not been interfered with, clots will be found adhering to the edges or to the wounded vessels. The edges of the wound will be found everted, and the muscular and cellular tissues of the wound effused with blood. If the wound has been inflicted a few hours before death, its edges will be found more or less swollen, and if of small extent, its lips will be loosely agglutinated.

If the wound has been inflicted some days it will exhibit either signs of repair or of destruction of the tissues. These may have healed by first intention or sometimes under cover of a scab, or by means of granulation which involves suppuration, and which is the common method of healing in wounds exposed to the air, where there is much injury to the parts. The appearance of these signs of repair is of course proof positive that the injury was inflicted before death. The earliest period at which pus will be found is from eighteen to twenty-four hours.

If the wound was inflicted after death its edges will usually be found in close apposition ; there will be comparatively little effusion of blood, and there will be very little or no coagula around the wound, and of course no evidence of repair or other vital reactions.

Contusions and contused wounds, which for convenience will be considered together, may give rise to effects of three sorts, namely, concussion, contusion, and disorganization.

As respects concussion, this effect will depend upon the degree of the impulse, and the nature and form of the part struck. The results of concussion will differ according to the part of the nervous system which is the seat of the injury; thus, concussion of the brain may cause momentary insensibility or death. Concussion of the spine may produce more or less injury to the parts receiving their nervous supply from the injured portion of the spinal cord. A stroke over the epigastrium or præcordia may by arresting the heart's action prove immediately fatal; where the concussion proves immediately fatal, death occurs by syncope, and not unfrequently, except in the case of the concussion of the brain, leaves no traces of the mode of its production in the interior of the body. In cases of concussion of the brain, however, there will usually be some local injury to the scalp; whether such fatal injuries will always produce a visible injury to the brain is a matter upon which surgical authorities are in conflict. In a case of concussion of the spine immediately fatal, there will usually be found traces of the injury, such as ruptures of ligaments, fractures or dislocations of the vertebræ, extravasation of blood outside or inside the cord, etc. Where, however, the concussion of the spine was not fatal, especially in cases of railway accidents, there is a great conflict of opinion as to whether the existence of real injury can be with certainty diagnosticated from malingering.

The term "contusion" is usually restricted to injuries unaccompanied by external wounds. A contusion strictly so-called involves the application to the body of sudden pressure accompanied with concussion, or of pressure continued for some time. It may be effected without rupturing the capillaries of the part struck, or it may involve the rupture of the small vessels. In the first place the contusion may not manifest itself by any striking phenomena; the part struck is painful; the meshes of the tissue of the skin have been compressed, and after some minutes the part swells slightly and reddens, the redness and swelling disappearing in from twenty-four to thirty-six hours, leaving no trace of the injury; but if at the time of the receipt of such contusion death supervenes from some other cause, the part struck undergoes by evaporation a loss of its fluids, and the skin becomes dried, brown, and hard, presenting much the appearance of parchment; which last effect, however, may be as readily produced on the dead as on the living, and will be further considered hereafter.

Where the stroke has involved the rupture of the capillaries the contusion will be accompanied by ecchymosis or infiltration of blood into the areolar tissue of the part. When such ecchymosis is superficial, as is usually the case, the skin soon becomes of a blackish or deep violet color; later this color succeeds to a bluish color; this in turn becomes green; the green becomes yellow; and the yellow is finally succeeded by the natural color of the skin, — these phenomena occupying a space of several days. As a general rule, the blue color appears about the second day; the green from the fifth to the sixth; the yellow from the seventh to the eighth; and the color entirely disappears from the tenth to the twelfth day, or even later. Ecchymosis may not involve the skin at all, but may be confined to the subcutaneous areolar tissue, in which case the discoloration of the skin will not appear till from thirty-four to thirty-six hours, or even later. Ogston describes a case in which ecchymosis did not appear till the expiration of four days after the infliction of the blows. In some cases where the ecchymosis exists deep among the muscles of the limb, it will not be noticed at all till at the end of forty, fifty, or sixty days, when irregular yellowish-green or bluish spots appear over the injured part. It should be remembered that the ecchymosis does not constantly appear at the situation the seat of the contusion would indicate; and that it may sometimes proceed from other causes than violence, such as scurvy, purpura hæmorrhagica, strong muscular effort, etc.

It is often important to determine whether contusions have been inflicted before or after death. When inflicted two or three hours before death the change of color produced cannot be confounded with any ordinary post-mortem appearance; and besides, the swelling and extravasation of blood into the cellular tissue will ordinarily afford conclusive evidence of the ante-mortem nature of the injury; but where the strokes are inflicted shortly after death there is more difficulty, as it has been proved by experiment that some strokes inflicted shortly after death produce marks which, so far as color is concerned, do not differ from the effects of blows during the last moments of life. Swelling of the parts and coagulation of the blood effused into the subjacent cellular tissue, with the incorporation of blood with the whole thickness of the true skin, rendering it black instead of white and increasing its firmness and resistance, may be considered as strong if not conclusive evidence that the blows were inflicted during life.

The co-existence of a wound with the contusion does not much affect the character of the latter injury; in this case, however, there may be hemorrhage in the surrounding tissues; much blood rarely escapes outwardly. Wounds of this class, unless over the cranium, are characterized by uneven and irregular edges, and less acute angles, and are in general too characteristic to require description. The disorganization of the parts struck, which may be one of the consequences when struck with a blunt body, and which is termed "attrition," differs both from contusion and laceration in that here the structure is more or less completely destroyed and broken down; attrition is necessarily followed by ecchymosis, — the broken-down tissue forming a cavity for the effused blood, which presents the characteristics of a tumor with fluctuation.

The most serious injuries resulting from contusions are lacerations of internal organs, and ruptures of blood-vessels and other deep-seated parts; such injuries, however, rarely follow common assaults with the fist or even with sticks or stones, unless the blows are inflicted about the head. The serious injuries referred to more commonly arise from falls from a height, the fall of heavy bodies upon the person, or from railway accidents. Ruptures of viscera are more common in the case of those organs which are naturally voluminous and easily displaced, such as the liver, kidneys, and spleen; and are more rare in the case of the heart, lungs, brain, bladder, and alimentary canal. Internal arteries and veins, when healthy, except those within the cranium and spinal canal, are not commonly injured by outward shocks. It must be remembered that ruptures of the heart are occasionally produced in certain diseased states of the organ, such as attenuation and aneurism of its walls, and it must also be remembered that spontaneous ruptures of the heart sometimes occur which cannot be certainly referred either to injury or disease. Ruptures of the cerebral mass from violence are believed to be extremely rare; in drunken quarrels and fights spontaneous ruptures of the cerebral vessels are not uncommon, and ruptures of the larger blood-vessels within other cavities of the body are not unusual consequences of heavy falls. The larger veins may also be ruptured from violence, and may burst suddenly when weakened by previous disease. For details of these interesting cases the student is referred to larger standard works on medical jurisprudence, the limits assigned to this work preventing their full discussion here.

Lacerated Wounds. Lacerated wounds, which are produced by tearing instead of cutting, are attended with complete separation of the edges, and there is always more or less thickening of their margins from the bruising of their edges. In this respect they somewhat resemble contused wounds, from which, however, they may usually be distinguished by the shreddiness and irregularity of their margins. Such wounds are remarkable for the want of correspondence between the quantity of blood lost and the importance and vascularity of the injured parts. They are frequently attended with marks of contusion, and there may be clotted blood effused in their vicinity. It is only when these appearances are present that a confident opinion can be given as to their having been produced during life.

Gun-shot wounds may be considered in this connection. These wounds partake of the character of contused wounds, with more or less laceration, and occasionally exhibit the appearance of burns. They bleed sparingly, if at all; their margins are round and thickened; the bottom of the wound is reddish brown, the surrounding parts ecchymosed and occasionally blackened. In the living they are usually attended with more or less insensibility of the parts struck, and the following inflammation is usually extensive and severe. When the projectile has passed through a portion of the body, as a general thing the entrance wound will be smaller than the exit wound, — the edges of the entrance will be depressed and contused; the edges of the exit projecting and torn. The entrance wound will be dry and dark-colored; the exit wound raw and bloody. There will be loss of substance in the former, but none in the latter as a general rule. Such is the description usually given of gun-shot wounds by writers upon medical jurisprudence and surgery. It should be remembered that substances of very low degree of density, such as plugs of tallow, light wood, cork, wads of paper, and the like, will, if sufficient velocity be impressed upon them, serve to inflict penetrating wounds into the denser parts of the body, having all the characteristics of wounds inflicted by projectiles. The effects produced by bullets, however, vary with the shape and speed of the projectile. Where the wound is produced by a rifle-ball at full speed, the above description of the wound is believed to be correct. As the distance from the weapon increases and the velocity of the ball diminishes, the wound of entrance will become less circular and regular, and larger and more contused, — the wound sometimes consisting of three triangular flaps, which on lifting up can be made to meet at their apices in the centre of the opening. Bullets at full speed perforate or penetrate; at lessened speed, crush and lacerate. When a discharge takes place very near the body the injured tissues will be more or less scorched, blackened and studded with grains of powder; and the entrance wound larger, ragged, and excavated. With the rifle-ball the course is more frequently direct than with that from the old smooth-bore musket, although tortuous courses are even now, though less frequently, met with. Modern projectiles are said to cause less severe injury of the soft parts than the old spherical balls. The injury imparted to soft parts by what are called spent balls and by ricochet shots are the most destructive; and larger projectiles rolling over the surface of the part or moving at a low rate of speed, possess a force which will often crush all parts with which they happen to come in contact. The character of the wound may also be modified somewhat by the hardness of the ball, and its change of shape in consequence of striking against a large bone. In such cases the size of the exit wound will be very much increased. The wound produced by the so-called "express bullet," in which the ball is light, has a concavity at its point, and is driven by a very large charge of powder, possesses the characteristic last mentioned in a most remarkable degree, even where the ball has not come in contact with any bone. An express ball after penetrating the body will usually expand and fly into many pieces, greatly lacerating the internal organs and producing great shock, much greater than if the ball had passed entirely through the body in its original shape. The medical jurist will, however, rarely have occasion to give an opinion upon wounds produced by this sort of projectile.

A charge of fine shot fired at a short distance will often produce nearly the same effect as that of a large solid ball; even the explosion of powder alone without ball or wadding, at a very short distance, is capable of producing very severe injuries or even death. It should be remembered, however, that a loaded pistol fired off with the muzzle firmly pressed against the body will probably burst or recoil without seriously injuring the person against whom it is fired. The statement made at the beginning of this subject, that the entrance wound is commonly smaller than the exit wound, is denied by some authors, and perhaps under some circumstances may require qualification; but in our judgment it is the better opinion.

The opening made by a ball in penetrating the clothes is always smaller than that made in the skin below them.

In this connection the question is sometimes raised as

to whether it is possible to determine from inspection of
the weapon the period which has elapsed since it was
last discharged. It has been found that when the combustion of the powder has been imperfect, the inside of
the barrel of the weapon near its muzzle is either found
blackened by a coating of charcoal and sulphide of potassium shortly after the discharge, — or where the combustion has been perfect, whitened by a crust of sulphate
and carbonate of potash, — while after an interval of
some days, varying with the amount of moisture in the
atmosphere, the mixed residue of charcoal and sulphide
of potassium has become converted into sulphate of
potash, which after a little longer interval has been
found to contain peroxide of iron. These results, however, might vary with the state of the weather, nature
of the powder, and the completeness of the combustion
to such an extent that in our judgment very little reliance can in most cases be placed on any conclusions
drawn from the appearance of the weapon.

Any one of the above previously described wounds
may be also a poisoned wound; but wounds of this description will rarely come under the notice of the medical jurist.

Burns and Scalds may be conveniently considered
together. A burn is caused by the application of concentrated dry heat to the body; a scald by the application of a hot or boiling liquid. As a rule, scalds are
less severe accidents than burns, because water, being
the ordinary fluid through which the scald is produced,
is never hotter than 212° Fahrenheit; yet when any
other chemical compound is the scalding medium, the
effect may be as bad as of the worst burns.

Molten metals produce burns which can hardly be distinguished from those caused by solid bodies. Boiling oils produce burns as severe in their general characteristics — and so far as destruction of parts is concerned, in their effects — as hot solids or melted metals.

Boiling water may produce merely an inflammatory redness; in severe scalds, however, the skin is commonly soddened, desiccated, and of an ashy-gray color, scarcely distinguishable from slight burns from other causes. Boiling water never produces, however, blackening of the cuticle nor charring nor desiccation of the parts.

Burns from acids are more properly considered in another connection. It may be stated here, however, that nitric or hydro-chloric acid stains are yellow, while sulphuric acid stains are brown, and that the eschars are soft and not hard as in ordinary burns; such burns are not surrounded by reddened skin, as in the case of those produced by heat.

According to Dupuytren there are six degrees of burns: —

In the first degree there is mere redness and tenderness of the surface, and after a few hours these symptoms may abate, with possibly some desquamation of the cuticle.

In the second degree there is inflammation, manifesting its presence by the effusion of serum beneath the cuticle, forming a blister.

In the third degree the superficial layer of the true skin is destroyed, the surface appearing of a gray, yellowish, or brown color, not painful unless roughly handled. The vesicles that exist contain a blood-stained or brown fluid. The papillæ of the skin with its nerves are first destroyed, but in the course of a day or so the dead surface is shed and the nerves exposed, when the pain becomes very severe.

In the fourth degree the whole thickness of the skin is destroyed, with more or less of the subcutaneous cellular tissue, the parts being converted into a hard, dry, and insensible eschar mottled with blood. The skin surrounding the eschars may be blistered, but where it comes in contact with the injured part it will be drawn into folds from the contraction, owing to the drying of the burnt integument.

In the fifth degree the skin with the deeper parts are involved, — a black, brittle, charred mass taking the place of healthy tissue.

In the sixth degree the whole thickness of the limb is carbonized.

Death following burns may take place in two different ways: from the depression of the nervous system, owing to the number of cutaneous nerves affected; or later, from the inflammatory reaction involved, extensive suppuration, and hectic fever.

Medico-legal questions in this connection usually arise where fire has been subsequently applied to the corpse to conceal murder. According to Christison, the most immediate effect of the application of heat to the living body is a blush of redness around the burnt part, removable by gentle pressure, disappearing in no long time, and not permanent after death. Following this almost immediately, is a narrow line of deep redness separated from the burnt part by a stripe of dead whiteness (bounded towards the white stripe by an abrupt line of demarcation), passing at its outer edges by different degrees into the diffused blush, but not capable of being removed like it by moderate pressure. The phenomenon which follows these is the appearance of blisters, which, when the agent is a scalding fluid, generally appear in a very few minutes in the living, or may be delayed for hours when the scalds are extensive, as in young children; while, when the agent is an incandescent body, this appearance is not of such an invariable occurrence, though often observed very soon after an ordinary burn, caused by the clothes catching fire.

· A line of redness near the burn, not removable by pressure, and blisters filled with serum are considered by Christison as certain signs that the burn was inflicted during life. The absence of these appearances, however, is said by Dr. Taylor not to point with certainty to the opposite conclusion. Dr. Ogston agrees with the above conclusions of Christison and Taylor, with the qualification that we must take into account not only the occasional failure, under certain circumstances, of vesication after vital burns, but also the non-occurrence in some instances of the redness of the burnt part; and in the second place, that vesication without accompanying redness on a dead body, would not authorize the conclusion that the burn had been caused during life, as such blisters are met with from pemphigus in the living, and in the corpse from the progress of putrefaction.

Where a body is more or less completely consumed, the sex may sometimes be distinguished by the pelvis, and the age from the bones or teeth. In case of death from burns it frequently happens that there are no characteristic post-mortem appearances; fluid in more than

usual quantity will frequently be found, however, in the
ventricles and at the base of the brain; also effusions
into the serous cavities. The bronchial tubes and lungs
generally are usually congested, as well as the stomach
and alimentary canal. Perforating ulcers of the duodenum are common, especially in children and young
people; the heart is sometimes found empty, but more
often the right side is full and the left empty. The
brain, liver, kidneys, and pelvic organs are frequently
congested.

Fractures. The subject of fractures may be conveniently considered in this connection. It should be remembered that they do not always arise from external
violence, but are sometimes caused by violent muscular
action, or in diseased conditions of the bone by ordinary
muscular action. Such injuries sometimes occur spontaneously in the *foetus in utero* and are sometimes occasioned during delivery; it also occasionally happens
that blows or falls of no great severity will cause fractures of a severe or fatal kind, only admitting of explanation by attention to the surrounding circumstances.

As to whether the fractures were produced during
life, the only test which can be relied upon with confidence is the commencement of the reparative process;
the presence of coagulated blood between the ends of
the fractured bones is not conclusive.

In rare cases the degree of violence received, or the
character of the instrument by which it has been inflicted, can be determined from inspection of the fractured bone.

The subject of blood stains will be considered in
another chapter.

The prognosis of wounds is a subject which will occasionally demand the attention of medical jurists.

1. **Prognosis of injuries to the nervous system.** Wounds of the head are important chiefly from their liability to disturb the functions of the brain or to involve it in active disease. When strictly local they heal without much trouble, but are liable to be complicated with diffuse abscess, erysipelas or irritative fever. When the irritation arising from such wounds is considerable, the brain and its membranes may sympathize, or the party be attacked with tetanus, and in this way an injury apparently inconsiderable may prove unexpectedly fatal. Contusions of the head are generally not dangerous if unattended with symptoms denoting concussion. Punctures or any simple fractures of the bones of the head, when the brain is not injured, are not in general followed by bad consequences, although a guarded prognosis should be given in every such case. Effusion of blood within the cranium is a very common cause of death from violence producing concussion of the brain, even when there is no external mark of injury on the head. The question in such cases arises whether the effusion is due to violence or to disease; such effusions, when spontaneous, are most usual in the substance of the brain, while if due to violence they are commonly on the surface or between the brain and the skull.

Injuries to the spinal cord which wound, divide, compress, or disorganize any part of it, generally prove fatal either immediately or after a longer or shorter period. Injuries to the medulla oblongata are instantly fatal in consequence of the immediate cessation of res-

piration and circulation; it is important, therefore, in all
such cases that the spine be carefully examined after
death.

2. **Injuries to the circulatory system** are common
causes of death, which may arise either from exhaustion,
due to hemorrhage or otherwise, or by the pressure of
the effused blood impeding the functions of the vital
organs, such as the brain, spinal cord, heart, or lungs.
The amount of blood which may be lost without destroying life varies with the different states of the constitution, and the habits of the party. The proofs of death
from hemorrhage are such as indicate that the wounds
were inflicted during life, the absence of blood in the
larger vessels and important viscera, and the healthy
state of the principal organs of the body.

Life may be immediately destroyed by the admission
of air into the veins through a wound.

Wounds of the neck severing the large vessels, in
the absence of surgical assistance are immediately fatal;
and even with such assistance they are usually fatal.

Wounds of the chest most frequently cause death
by syncope from hemorrhage, but a less effusion may
destroy life by compression of the heart or lungs.
Ruptures of the heart or larger vessels within the
chest may occur either from pressure or from a blow,
without any appearance of external injury. They may
also occur spontaneously from disease.

Wounds of the organs of generation are occasionally
fatal when no large vessels have been divided, such as
incisions on the inner side of the labia, nymphæ, or
vagina. There are recorded instances in which a nearly
fatal hemorrhage has been caused by coitus.

3. The most important **injuries of the respiratory system** are of the chest and lungs. Contusions, and fractures of the ribs, wounds of the pleurae and lungs, may prove fatal from shock or asphyxia. In penetrating wounds of the chest, the entrance of the air may cause collapse of the lungs and death by asphyxia. Wounds laying open the larynx and trachea, while not necessarily fatal, may cause death by asphyxia from the blood obstructing the air passages, or from fluid swallowed passing into the larynx and causing suffocation.

4. **Injuries of the abdomen** may prove fatal by shock, hemorrhage, inflammation, or by interference with the nutrition of the body. It should be remembered that spontaneous ruptures of some parts of the bowels may sometimes occur and cause sudden death.

The prognosis of an injury may be modified by age, sex, and constitutional peculiarities; thus in some persons the bones are so thin and brittle as to be easily fractured by slight blows or falls. In what is known as the hemorrhagic diathesis, dangerous or even fatal hemorrhage may follow a trifling blow or cut.

In the medico-legal inspection where death has been caused by wounds, there are some points needing special attention. The nature of the wound, whether recently inflicted, whether inflicted during life or after death, should if possible be carefully determined. Any weapon found should be carefully compared both with the clothes and the external wound, and foreign bodies (if any) found in the wound should be carefully preserved. The length, breadth, and depth of the wound should be measured, and its situation and direction carefully noted. The probable manner of infliction and the degree of force and the weapon employed may often be ascertained from the nature and extent of the wound. The question as to whether death has been the result of accident, suicide, or homicide, may sometimes be determined from the above circumstances. The manner of dissecting the wound requires careful attention. For this purpose a circular incision should be carried around the wound, three or four inches from it and not interfering with it, and the integument and underlying structures successively dissected off, from the circumference to the centre; in this way the relations of the parts implicated, and the direction and extent of the wound may be accurately determined. The general dissection of the body may then be completed in accordance with directions given in a previous chapter.

CHAPTER IV.

By somatic death, or the death of the body as a whole, is meant the cessation of the vital functions, and of the general renewal of tissue consequent on such cessation.

By molecular death, which may be either partial or complete, is meant the death of a part, tissue, or organ, without the general stoppage of the circulation. The part thus affected becomes obedient to the operation of the ordinary chemical and physical agencies governing the inorganic molecule.

The **signs of death** are —

1. The entire and continuous cessation of the heart's action. This cannot be certainly determined by the mere absence of pulsation at the wrist. In order to decide this question an examination both by auscultation and palpation, in a perfectly quiet room, are necessary; and this examination should include not merely the regions of the heart proper, but of the chest generally, and a positive conclusion should not be announced until after careful auscultation for two or three hours, at intervals of fifteen minutes; for recovery has been known to take place after the heart has apparently

4

ceased beating for the period of fifteen minutes. The
auscultation is better made with a good stethoscope
than with the naked ear. In addition to palpation and
auscultation, fine chest movements may be detected by
certain mechanical tests, such as sticking needles with
little paper flags on their blunt ends into the skin, or
placing small pieces of cotton wool drawn into finely
pointed cones about two inches in length, over the
region of the heart, and of the great vessels of the neck;
the room meanwhile must be kept perfectly still and
free from draughts.

Another test is the placing of a ligature tightly
around a limb of the body, such as the finger, or
around the lobe of the ear. The part beyond the con-
striction will, if the person is alive, become bright red;
the tint gradually increasing in depth until it finally
assumes a uniform bluish-red color. At the spot
where the ligature is applied a narrow white ring will
become visible. There will be no change if the person
is dead.

Again, if the person has been dead some hours, no
blood will flow upon scarification and application of a
cupping-glass.

Again, if during life clean and bright needles are
thrust into the muscles, the steel will quickly oxidize
and tarnish; but after death the needles may remain in
the flesh an hour without such oxidation.

A superficial artery may also be cut down upon and
its color and contents ascertained. After death the
arteries are pale or yellowish and empty of blood,
while during life they pulsate, and have the color of
the surrounding tissues.

If during life a little ammonia solution be subcutane-
ously injected, a port-wine congestion is set up in the
surrounding tissues, but no such redness results when
the operation is performed upon the dead body; at
least, not if the body has been dead for some hours.

2. Another sign of death is the entire and continuous
cessation of respiration, which may be determined —

a. By holding a cold looking-glass over the mouth
and nose, when if respiration is present the moisture
exhaled will condense upon the mirror; the absence,
however, of such condensation is not a conclusive
proof of death:

b. By suspending a feather or other light body near
the mouth and nose; here again, the caution above
given is applicable.

c. By standing a glass of water or mercury on the
naked chest, whereby the slightest motion will become
perceptible.

With reference to these last three tests, it should be
remembered that it is consistent with life that for a
short time respiration may be practically imperceptible;
it is only the entire and continuous cessation of respiration which is indicative of death.

3. Insensibility and the loss of voluntary motion,
while they frequently occur without resulting in death,
should be noticed in this connection. Such insensibility and inability to move may arise from asphyxia,
syncope, apoplexy, trance, catalepsy, the mesmeric state,
and cases of long and persistent sleep, without the
person's being dead.

By way of caution it may be remarked that a physician is not justified in certifying the death of the

person unless the majority of the signs of death are well marked; he should never be satisfied of the fact of death from one or two appearances merely.

4. There are certain minor signs which may be here considered —

a. Dry heat may be applied to the skin; if this produces a blister containing a serum rich in albumen, while the true skin after the cuticle has been removed, presents a reddened appearance, and more especially if after a short interval a deeply injected red line forms around the blister, this is absolute evidence of the vitality of the part to which the heat is applied, and consequently strong confirmatory evidence of the life of the person; but if the blister contains merely air or a little non-albuminous serum, and the true skin after the removal of the cuticle appears dry and glazed and no red line appears around the blister, it is certain that the part so treated is dead, and the evidence is therefore strong that the person himself is dead.

b. Caustic may be applied to the skin; if the skin is living the eschar is of a black or reddish-brown color; but if dead either no eschar is produced or the skin turns yellow and transparent.

c. Another change prior to decomposition is what is known as the facies Hippocratica, which is characterized by sinking of the eyes, hollowness of the temples, sharpness of the nose and chin, dryness and harshness of the forehead, sallowness of the countenance, flaccidity and paleness of the lips, all which precede death and continue to be recognizable after dissolution. Another almost constant appearance in the dead body is a more or less flexed state of the fingers and thumbs,

the latter from this cause being sometimes bent across
the palms and the fingers closed on them.

5. There are certain changes in and about the eye
which are indicative of death; thus —

a. The iris becomes insensible to light, or, in other
words, the pupil does not change its size with increase
or diminution of light thrown upon it. This symptom,
however, occurs in many cases of disease and in long-
persistent sleep, and may also be produced by the action
of certain drugs, such as atropine, eserine, etc. After
death the iris becomes more or less flaccid, and external
pressure may permanently affect the normal roundness
of the pupil.

b. Again, the cornea becomes insensible; but this also
occurs during certain stages of epileptic fits and in
certain cerebral injuries.

The eye loses its lustre and the cornea its trans-
parency, as a rule, soon after death; afterwards its
tension becomes lessened and the cornea wrinkled and
flaccid from the absorption of the aqueous humor. But
here again, the eye may lose its lustre during life, and
sometimes its lustre is preserved a long time after death,
as after poisoning with the oxides of carbon, cyanogen
and its compounds, etc. The same condition has also
been observed after death from apoplexy.

c. Soon after death the conjunctiva shows gray, cloudy
discolorations, which rapidly become black.

d. Loss of tonicity and of the elastic resistance of the
eye usually occurs in about twelve or fourteen hours
after death; but this, again, may occur during life.

e. Where the cornea is clear enough to allow ophthalmoscopic examination, the change of the fundus from

the yellowish-red color of the living to a yellowish-white hue is a strong indication of death. After death, also, the retinal veins show a beaded condition, due to the liberation of gases disengaged from the blood.

6. Changes in the temperature of the body. The healthy living body has normally a temperature of about 98.6° Fahrenheit, or 37° Centigrade, which after death gradually falls to the temperature of the surrounding medium. The post-mortem cooling derives its importance from the fact that this loss of heat is progressive, so that the temperature is many times not merely a sign of death, but an indication of the length of time the body has been dead. The temperature should be taken at regular intervals with a reliable thermometer, the corrections of which are known. Both the external and internal temperature should be taken, — the external temperature in the axilla, and the internal in the mouth or rectum.

The time within which a body ordinarily becomes cold is stated by Casper to vary from eight to twelve hours. Dr. Tidy thinks this period too short, and fixes the time at from fifteen to twenty hours after death. Other writers make the time even longer, — as from eighteen to twenty-four hours.

In certain cases of disease or where the body has been freely exposed to the air, draughts, etc., the cooling process may be completed within a much shorter time, even in four or five hours; while in certain other exceptional diseases or under unfavorable conditions for cooling, even forty-eight or seventy-two hours may elapse before the body is cold. In certain cases, such as death from yellow fever, cholera, Bright's disease, abscess of the liver and other abdominal affections, rheumatic fever, small-pox, tetanus, and injuries to the nervous system generally, there may be a post-mortem elevation of internal temperature amounting sometimes to as much as nine degrees Fahrenheit. The following table contains the results of the researches of Drs. Taylor and Wilks on external temperature, the thermometer being placed upon the skin of the abdomen.

	2 to 3 Hours after Death.		4 to 6 Hours after Death.		6 to 8 Hours after Death.		12 Hours or more after Death.	
Number of Observations.	76		49		29		35	
	F.	C.	F.	C.	F.	C.	F.	C.
Maximum temperature of the body .	94°	34.4°	86°	30.0°	80°	26.6°	79°	26.1°
Minimum temperature	60	15.5	62	16.6	60	15.5	56	13.3
Average temperature	77	25.0	74	23.3	70	21.1	69	20.5

Of internal temperatures, Taylor and Wilks record cases of 76° F. (24.45° C.) seventeen and eighteen hours after death, and of 85° F. (29.45° C.) ten hours after death.

The following table of external temperatures records the results of 135 observations on the bodies of persons who had died from various diseases. The temperatures were taken by placing the thermometer in the axilla:

Temperature of Body after Death.	2 to 4 Hours.		4 to 6 Hours.		6 to 8 Hours.		8 to 12 Hours or more.	
	F.	C.	F.	C.	F.	C.	F.	C.
Maximum	109.4°	43.0°	98.2°	36.8°	95.3°	35.2°	100.4°	37.8°
Minimum	89.6	32.0	80.6	27.0	70.5	21.4	62.6	17.0
Average	96.9	36.1	90.2	32.3	81.7	27.6	77.9	25.5

Authorities differ very much among themselves as to the rate of post-mortem cooling. The important fact to bear in mind in this connection is, that the rate of cooling is not uniform; notwithstanding the post-mortem rise of temperature, it is on the whole during the earlier hours after death that the most rapid cooling occurs; during the later hours the loss per hour becomes exceedingly trifling.

The time occupied by the cooling process may be shortened in death from wasting diseases, after great losses of blood, where the body is exposed to air and cold draughts in a more or less uncovered state, on the floor or on other good conducting surface. It will cool more rapidly in a large well-ventilated room than a small close one. Bodies of children or old persons cool more rapidly than those of adults. Bodies of lean people cool more rapidly than those of fat. The time of cooling in death from drowning may be shortened by the temperature of the water. Again, the time of cooling may be lengthened in acute diseases, generally in sudden death and in cases of asphyxia, except drowning. Warm clothing, non-exposure to draughts, preservation in a small warm room, being well covered or on a non-conducting material, will also prolong this process.

7. Changes in the muscles and the general condition of the body after death.

a. Shortly after death the muscles become flaccid, the jaw drops, the eyelids lose their tonicity, the joints become flexible and the limbs flabby. During this period, however, the flabby muscles are capable of contracting under appropriate stimuli, such as interrupted electric currents, blows, etc. This stage may occasionally last only a few minutes, or even be non-existent; but it more commonly lasts about three hours. There are no well-attested cases, however, where this stage has been prolonged beyond twenty-four hours. It is to be observed that the contractility of a muscle by electrical and other stimuli is no certain test of life; and that the non-contractility of a muscle by such stimuli is no certain test of death.

b. At the end of the first stage of flaccidity and irritability follows what is termed the stage of cadaveric rigidity, or *rigor mortis*, by which is meant rigidity of the muscles accompanied by stiffness of the joints and limbs. As soon as the elasticity and muscular irritability cease, *rigor mortis* commences. During this stage the muscles retain the precise position they occupied at the time rigidity supervened. This phenomenon is common both to voluntary and involuntary muscles, and is altogether independent of the nervous system, of the presence of the air, and of temperature, although the early supervention of *rigor mortis* may have some influence in quickly lowering the surface temperature.

The true cause of *rigor mortis* is believed to be the coagulation of the myosin or muscle fibrine, the albuminous principle of the muscular tissue. As putrefaction proceeds, ammonia is developed, the coagulated myosin is dissolved, and *rigor mortis* disappears. The reaction of the living muscle at rest is faintly alkaline; contracting muscle possesses a faintly acid reaction. During *rigor mortis* the muscle exhibits a well-marked acid reaction, but when *rigor mortis* has passed away the muscles exhibit a well-marked alkaline reaction.

Rigor mortis in the voluntary muscles usually commences at the third or fourth hour after death, and is usually complete at about the fifth or sixth hour; it comes on sooner in the involuntary muscles than in the voluntary.

Rigor mortis may be late in appearing in cases of sudden death in muscular and well-developed subjects; under such circumstances, when the muscles have not been previously fatigued or the body weakened by disease, rigidity may not set in under twelve hours, or even longer. Again, if the body is exposed to cold *rigor mortis* is often delayed; there is, however, no authenticated case where it has been delayed beyond twenty-four hours.

On the other hand it may appear very soon after death, especially in certain parts of the body. The eyelids may become rigid within five minutes of death, or sometimes, it is said, before the heart has ceased to beat; rigidity often sets in very rapidly in the facial muscles. There are numerous recorded cases where rigidity has set in while the body has been warm, and a case of death from typhoid fever is recorded where rigidity commenced while the heart was still beating, and within three minutes after respiration had ceased.

Living contraction may pass at once into *rigor mortis* without any appreciable intermediate state of muscular flaccidity, and these cases are confined to no one special mode of death. It has been commonly observed that where immediate rigidity occurs the period just preceding death has been one of great fatigue and physical exhaustion.

When the last attitude of life is maintained after

death by *rigor mortis,* considerable light may be thrown upon the question whether the case is one of homicide or suicide. In such cases the position of the dead body, its relation to the surface on which it rests, the position of the weapon or other thing grasped by the hands should immediately be carefully noted. If a weapon be found loosely held in the hands of the deceased no conclusion of value can be deduced as to the question of suicide or homicide; but if the weapon be found firmly grasped by the deceased, suicide rather than homicide is indicated.

As to the order after death in which the various parts of the body are affected by *rigor mortis,* while there are some minor differences among the authorities, they in the main agree. The eyelids appear to be first attacked; after that it passes from above downwards, beginning, according to Casper, on the back of the neck and lower jaw, passing thence to the facial muscles, the front of the neck, the chest and the upper extremities, and last of all the lower extremities. It usually passes off in the same order and once gone never returns, the body becoming as flexible as it formerly was.

Rigor mortis sometimes passes off very rapidly, so rapidly that it has been reported never to have occurred; and again it sometimes lasts a long time. The length of time during which it lasts is variously stated, — at from sixteen to twenty-four hours (Taylor), and twenty-four to thirty-six hours in summer, and thirty-six to forty-eight in winter (Tidy). The time of its appearance and the duration of. its continuance are modified by age, temperature, and atmospheric conditions, condition of the muscular system, and the mode of death. It is less

marked in the bodies of middle-aged persons (unless the
subject be very muscular) than in the old, where it is
most complete.   In infants it usually sets in very
rapidly.   A low temperature and dry air favor a long
continuance of rigidity.   In all cases of *rigor mortis*
two points require special consideration, — the muscular
development of the subject, and the extent of exhaustion
and fatigue preceding death.   Rigidity often lasts a long
time after violent death; after sudden death in a muscular subject it often continues fourteen days or even
longer.   Where death results from a lingering disease,
accompanied by great prostration, or from violence preceded by intense physical fatigue, *rigor mortis* sets in
speedily and disappears quickly; it may even be so
slight as to be overlooked.   Paralyzed limbs are subject
to *rigor mortis*; in cases of poisoning, rigidity as a rule
sets in late and lasts long; in poisoning, the primary
question as regards rigidity is not so much the action
of the poison as the intensity of exhaustion which has
succeeded death.   Rigidity usually continues long in
cases of habitual drunkards.

It is said that after death from small-pox, acute rheumatism, tetanus, meningitis, abdominal diseases, pyemia,
and the like, bodies become rigid rapidly and remain so
a long time.

Rigor mortis may be distinguished from other forms
of rigidity, such as may occur during life, by forcibly
bending the joint; when a joint stiff from post-mortem
rigidity is forcibly bent the rigidity passes away and
does not return, provided *rigor mortis* is completely established.   If it has not completely set in when the
limb is bent, a certain but less marked stiffness may

return. When forcibly bent the limb affected by *rigor mortis* does not return of its own accord to its original position; but in the rigidity of hysteria, catalepsy, syncope, or that caused by the action of certain poisons, the stiffness is not destroyed by forcibly bending; as soon as the force is removed the limb at once returns to its original position. Post-mortem rigidity is also accompanied by a progressive loss of heat which is not characteristic of any disease.

· *c.* The third stage of the changes in the general condition of the body after death is that of putrefactive decomposition, which is a spontaneous change common to all nitrogenized organic bodies when exposed to air, whereby they become resolved into new and simpler products. It is accompanied by the evolution of gaseous compounds, for the most part of sulphur and phosphorus. This stage usually commences when rigidity ceases, although in exceptional cases rigidity and putrefaction may coexist. Generally, an advanced putrefaction is an infallible sign of death; partial putrefaction after some local injuries, such as the gangrene of a portion of the body, is not evidence of death.

Before considering the appearances due to putrefactive decomposition, cadaveric ecchymoses, variously called hypostases, cadaveric lividities, post-mortem stains, sugillations, vibices, should be considered, although chronologically they belong to the first stage of muscular flaccidity, and are not the result of putrefaction. By cadaveric ecchymoses are meant certain post-mortem stains closely resembling in their general appearance, bruises or contusions, and occurring both externally and internally on the lowest parts of the body.   · ·

They are both external and internal. External ecchymoses usually appear within the eight or ten hours
after death, while the body is warm and the blood
liquid. When the blood is coagulated and the body
cold their progress and formation cease. The blood
within a dead body coagulates in much the same manner as it does when withdrawn from the living body.
When drawn from the body during life or within the
first three or four hours after death, it coagulates almost
immediately upon its exposure to the air; but when
remaining in the body from six to ten hours more may
elapse before coagulation. The time required for coagulation, whether in or out of the body, depends upon the
quantity of fibrin in the blood. As this varies greatly
in different diseases, the formation of post-mortem ecchymoses varies according to the cause of death, — coming on quickly in acute inflammations, and slowly in other
diseases, where the quantity of the fibrin in the blood is
small, as in phthisis. These external ecchymoses, lividities, or hypostases may readily be distinguished from
appearances likely to be confounded with them, by
attention to the following criteria: —

First, their seat, the superficial layer of the true
skin; secondly, their extent, involving large portions
of the body, or to be met with in different parts of it
at the same time; thirdly, their circumference, which,
though irregular and slashed, terminates abruptly, and
not by gradual fading into the surrounding colorless
skin; fourthly, the entire absence of extravasated
blood at their site; and fifthly, by the absence of any
trace of contusion or ruffling at the part of the cuticle
or true skin.

If a contusion produced during life is incised, blood at once flows from the cut; but no effused or coagulated blood escapes upon incision of a post-mortem ecchymosis, although perhaps a few bloody points may be apparent. Internal ecchymoses or hypostases also occur in the dependent parts of the several viscera, and this fact will serve to distinguish them from redness due to disease; they are chiefly found in the posterior part of the brain, the posterior part of the spinal cord, the posterior part of the lungs, and the dependent parts of the stomach and intestines.

Very soon after death the coloring-matter of the bile oozes from the gall-bladder; and the contiguous parts of the stomach and intestines may become thereby stained and of a yellowish or greenish-yellow color.

Coming now to **putrefaction proper**, it may be observed that its progress varies considerably according to the character of the soil, temperature, humidity, dryness, etc., and whether the body is immersed in water, — putrefaction therein being slower than in the atmosphere, — and upon other circumstances. A temperature of sixty to ninety degrees Fahrenheit is the most favorable to the process of putrefaction; in water the process is very rapid when the fluid has a temperature of from sixty-four to sixty-eight degrees, but very slow if the temperature of the fluid is lower than this. Other things being equal, very young infants putrefy sooner than adults or old people, and females sooner than males. Mutilations, contusions, ecchymoses, and death from acute diseases hasten putrefaction. Copious hemorrhages before death retard it.

The gaseous products of decomposition in the early

stages of putrefaction consist principally of sulphuretted and carburetted hydrogen and ammonia, with variable proportions of carbonic oxide, phosphoretted hydrogen, nitrogen, and carbonic anhydride. After the fifth or sixth day the proportions of sulphuretted hydrogen and ammonia diminish, and carburetted hydrogen and carbonic oxide largely predominate.

The principal phenomena attending putrefactive decomposition are color changes and the development of gases of decomposition, with their consequences. The order in which the external phenomena of putrefaction occur in bodies exposed to the air, or buried, are thus stated, as average results only, by Dr. Tidy, whose rules are with some modifications quoted from Casper : —

From twenty-four to seventy-two hours after death, a light-green color appears about the centre of the abdomen; the eye-balls are soft and yield to external pressure.

From three to five days after death the green color of the abdomen becomes intensified and general, spreading, if the body is exposed to the air or buried in the ground, in the following order: genitals, breast, face, neck, superior and inferior extremities. If submerged in water the following is the order, the time of appearance being later than in case of air : face, neck, shoulders, sternum, abdomen, and legs.

From eight to ten days after death the color becomes more intense, the face and neck presenting a shade of reddish-green. The ramifications of the subcutaneous veins on the neck, breasts, and limbs, are very apparent; the patches congregate; gas begins to be developed in and to distend the abdomen and the hollow organs, and to form under the skin in the submucous and intermuscular tissues. The cornea falls in and becomes concave. The *sphincter ani* relaxes. The nails remain firm.

From fourteen to twenty-one days after death the color of the whole body becomes intensely green, with brownish-red or brownish-black patches; the body generally is bloated and appears large from the development of gas in the abdomen, thorax, scrotum, and the cellular tissues of the body generally. The swollen condition of the eyelids, lips, nose, and cheeks usually obliterates the features so as to prevent identification of the body. The epidermis peels off in patches, while in certain parts, especially the feet, it will be raised in blisters filled with a red or greenish fluid, the skin underneath frequently appearing blanched. The color of the iris is lost. The nails easily separate. The hair is loose.

From four to six months after death the thorax and abdomen burst; the sutures of the skull give way; the viscera become pulpy, or perhaps melt away, leaving the bones exposed. The bones of the extremities separate at the joints.

At an advanced stage the soft parts gradually disappear. Generally, the changes in a body enclosed in a coffin are similar to but slower than those which appear if the body is exposed to air. After periods varying from a few months to one and a half or two years the soft tissues of a body buried in a coffin usually become dry and brown, and the limbs and face covered with a soft white fungus. After a period of four years the viscera become so mixed together that it is difficult, if not impossible, to distinguish them. At later periods the soft parts as a rule, entirely disappear. The teeth,

5

bones, and hair are the most indestructible parts of the body, and may be found in perfect preservation after many years' burial. Children's bones, however, decay more rapidly than adults'.

Besides the distention of the body by the gases of decomposition already mentioned, they also cause blood displacements and fluid effusions, and sometimes movements of the body simulating vital acts. The pressure produced by the gas usually empties the heart and large blood-vessels, and forces the blood either into the superficial capillaries, or into the mucous or serous membranes, or into the vessels of the viscera, causing a diffused and intense redness of the skin and areolar tissue. In post-mortem redness the red color is limited to the course of the vessels, and may thus be distinguished from inflammatory redness, which is more widely spread. The pressure resulting from the development of gases explains the occurrence of the so-called post-mortem hemorrhages, where liquid blood oozes from a wound before putrefaction has commenced. The fluid effused into the serous cavities, especially the pleuræ and pericardium, is caused by the pressure on the overloaded capillaries forcing the serum, brownish-red in color and homogeneous in nature, through the lining walls of the vessels.

The following organs and parts putrefy rapidly, and as a rule in the order mentioned :—

1. The larynx and trachea. 2. The brain of infants (the adult brain putrefies much less rapidly). 3. The stomach. 4. The intestines. 5. The spleen. 6. The omentum and mesentery. 7. The liver, which in children putrefies sooner than in the case of adults. 8. The brain of adults.

The following organs putrefy slowly, and generally in the order mentioned : —

1. The heart. 2. The lungs. 3. The kidneys. 4. The bladder. 5. The gullet. 6. The pancreas. 7. The diaphragm. 8. The larger blood-vessels. 9. The uterus, which resists putrefaction longer than any of the other soft parts of the body.

As putrefaction advances in bodies exposed to the air, vermin make their appearance. Recent investigations tend to show that the time which has elapsed since death may be approximately determined by a skilled entomologist from the order of development of the insects in the body.

In exceptional cases the body becomes desiccated and mummified, as where the body is exposed to dry air and protected from moisture.

Under certain conditions the soft parts of the body may be converted into what is called adipocere, which is an ammoniacal soap, requiring for its formation fatty matter and nitrogenous matter capable by decomposition of yielding ammonia. It is formed more readily and abundantly in fat than in lean bodies, in children and in young people than in adults. Complete immersion of the body in running water favors its formation ; as does also burial in an overcrowded churchyard, in cess-pool soil, or in deep graves.

There is also an adipocere having lime and not ammonia for its base, which is commonly harder and whiter than ammonia adipocere.

Adipocere is a white or yellowish-white or brown, soapy body, of an offensive rancid odor, which, when perfectly dry, is white, hard, and brittle. It is lighter than water, and melts at about two hundred degrees Fahrenheit; it is very durable. The breast, cheeks and kidneys are the parts first to undergo this change, and later the muscles. The time required for its production is variously stated. Casper says that in his experience it is not formed to any extent under three or four months' submersion in water, or six months' burial in moist earth.

There are recorded instances in which adipocere has been produced by immersion in water, or in water-closet soil for a period of a month or six weeks; and there are several recorded cases of adult bodies being found partly adipoceratous after three or four months' submersion in water; and there are many recorded cases where the change has been almost complete after two or three years' burial or submersion. Partial saponification may therefore be expected under favorable conditions after three months' submersion in water or twelve months' burial in earth.

Modes and Causes of Death. Medical jurists describe three modes of death, namely: by **Syncope**, or death beginning at the heart; by **Coma**, or death beginning at the head; and by **Apnœa**, or as generally but improperly called, **Asphyxia**, or death beginning at the lungs.

Death by Syncope may depend upon two distinct causes: where there is a deficiency of blood but no deficiency of heart-power, which is called Anæmia; or where there is a deficiency of heart-power but no want of blood, which is called Asthenia. The symptoms of death by anæmia are a mortal paleness of the cheeks and lips, cold sweats, dimness of vision, dilated pupils, giddiness, and a weak, irregular, or fluttering pulse. There may be nausea and vomiting, restless movements of the limbs, transient delirium with frequent hallucinations of the sense of hearing, and flashes of light before the eyes. The breathing becomes irregular, with sighs and often gasping; there is often hiccough. Insensibility eventually sets in. Convulsions generally supervene and may be repeated once or twice before death. The heart, if examined soon after death, will be found contracted, and quite, or nearly empty.

In death by asthenia, the hands, feet, and surface generally become cold; circulation in the extremities is usually first arrested, so that the fingers, lips, nose, and ears become livid. The pulse becomes feeble and frequent, and muscular weakness extreme. The senses and intellect retain their full activity to the last; the clearness of intellect distinguishes collapse from concussion in which consciousness is temporarily lost. In asthenia the cavities of the heart after death are not contracted, but are more or less full of blood, or if empty, flabby and dilated.

In death by syncope proper, we have a combination of anæmia and asthenia. Sudden death from shock is probably the result of syncope, — from causes acting through the nervous system. The general post-mortem appearances after death from syncope are, that the right and left sides of the heart commonly contain an equal amount of blood. As a rule, the brain, lungs, and capillary system generally will be found in a normal condition.

In Coma, or death beginning at the head, there is

stupor more or less profound; external impressions are but feebly recognized; loss of sensibility and consciousness gradually becomes complete. The breathing becomes slow, irregular, and stertorous, and voluntary control of respiration is lost; respiration may, however, be imperfectly carried on for a time after consciousness has ceased. Finally, the chest ceases to expand, the blood is no longer aerated, and death ensues. The post-mortem appearances in the thorax may differ but slightly from those found in death beginning at the lungs. The arteries and left side of the heart are empty; the right side of the heart and lungs moderately full, but not so engorged as after death from apnœa. Possibly effusions may be found within the head.

In Apnœa or Asphyxia, where the death begins at the lungs, there is an intense struggle to breathe preceding unconsciousness, vertigo, loss of consciousness, relaxation of the sphincters, general convulsions, and death. As a general thing the right side of the heart, veins, capillaries, and viscera are engorged with blood, while the left side of the heart and arteries are comparatively empty.

In many deaths, a combination of these three methods will be found, and it is often difficult to say which is predominant. German medical jurists have accordingly adopted a fourth form of death, called comato-asphyxia, where the death begins both at the brain and lungs.

The Causes of Sudden Death are various. Without professing to give an exhaustive enumeration, there may be mentioned, excluding violence and poisoning, diseases of the heart, especially fatty and brown degeneration, angina pectoris, aortic regurgitation, rupture of the heart,

aneurism and thrombosis, effusions of blood in the brain
or its membranes, pulmonary apoplexy and hemato-
thorax, the rupture of visceral abscesses, ulcers of some
portion 'of the alimentary tube, extra-uterine fœtation,
rupture of the uterus, apoplexy of the ovary, peri- and
retro- uterine hæmaticele, rupture of the bladder or some
other important viscus, cholera, and the accidental
obstruction of the pharynx or glottis by foreign bodies.
Mental emotions sometimes cause sudden death.

CHAPTER V.

Asphyxia means the imperfect aeration or non-aeration of the blood from want of air. There are certain points of resemblance in death by drowning, strangulation, suffocation, and the action of poisonous gases, which may be conveniently considered at this point, reserving characteristic differences for separate treatment.

The general symptoms of death by asphyxia are lividity of the whole face and of the extremities, with convulsive movements, at first more or less voluntary. Consciousness is soon lost; involuntary and unconscious spasmodic clonic movements of the muscles and limbs follow; the veins become turgid; the pulse more and more feeble. There is often frothing at the mouth, the froth sometimes being tinged with blood; blood not unfrequently escapes from the nostrils, vagina, anus, and other mucous tracts. The urine, fæces, and semen may be discharged involuntarily; abortive efforts at respiration continue for a while, but finally cease, and after an interval the heart also ceases to beat.

Dr. Tidy makes four distinct periods in asphyxia:—

(1.) A period of intense sensible although ineffectual efforts to breathe.

(2.) A period of insensibility, with more or less irregular convulsive and involuntary spasms.

(3.) A period when, though life *seems* to be at an end, owing to the failure of respiration, it may be again resuscitated, because of the continuance of the heart's action.

(4.) Its termination in death after the cessation of the action of the heart.

The first two periods may occupy from three to five minutes; from the beginning to the end, ten minutes may be regarded as an outside limit.

The post-mortem appearances of death from asphyxia are lividity of the lips, extremities, and general surface; hypostases or post-mortem stainings; the engorgement of the veins over the entire body with dark blood (the arteries being for the most part empty). The lungs are usually congested. The right heart, vena cava, and pulmonary arteries are distended with dark fluid blood, while the left heart, aorta, and pulmonary veins are usually nearly empty. Extravasations of blood will be found on the mucous and serous membranes. The membranes of the brain will usually be found gorged with blood, and numerous bloody spots appear when the brain is sliced. Serum will be found extravasated into the serous cavities, and the mucous membranes will generally be found very turgid. *Rigor mortis* is said to set in slowly; the viscera will generally be found enlarged and congested with dark venous blood, while the blood itself will be found usually fluid and dark in color. In death, however, from the effects of carbonic oxide the blood will be found of a bright red color.

In hanging and other violent deaths the genital organs

are often turgid and erect, while in drowning the penis is often retracted and the scrotum shrunk.

Drowning. Distilled or rain water having a specific gravity of 1.000 at 60° Fah., and sea water about 1.028, while the human body has a specific gravity of from 1.08 to 1.10, slightly greater than that of any water, the naked human body placed in water has a slight tendency to sink. Some bodies, however, will not sink in sea water.

In death by drowning there are two distinct sets of phenomena. First, those due to suffocation, and occasionally syncope and other causes; and secondly, those due to prolonged immersion in water. Under ordinary circumstances, death from drowning results from asphyxia. Asphyxia commences in about two minutes; and five minutes after submersion, death ordinarily occurs. Although apnœa or asphyxia is the most common cause of death, death may result, or at least its occurrence be modified, by syncope due to shock, fright, drunkenness, hysteria, catalepsy, etc. Death may also result from exhaustion, or concussion, — caused by collision of the body with some hard object, as the bed of the river, rocks, etc., — from apoplexy, cramps, or in epileptics, from the sudden advent of an epileptic seizure.

According to Devergie, death results from true asphyxia in 25 per cent of cases of drowning, and in the remaining 75 per cent, from asphyxia in a more or less modified form. In all probability, therefore, the general post-mortem appearances will be those of asphyxia already described. The post-mortem appearances, however, will vary according to the length of time of submersion, whether the body was completely submerged, has risen after submersion, etc.

Among the external appearances in case of death by drowning, may be mentioned the following: —

The position of the body may constitute important evidence; thus, if a rope or piece of any substance is found tightly clutched in the hand, or if two persons are recovered clasped in each other's arms, the conclusion that death has resulted from drowning will be greatly strengthened.

Cutis anserina, or goose skin, is found in many cases, and although not pathognomonic of drowning, its presence is strongly suggestive of submersion of the body, either during life or soon after death.

The face in case of drowning is usually pale and placid, though sometimes rosy red. If, however, two or three days in summer, or eight or ten in winter, have elapsed before the body is recovered, or if after recovery it has been exposed for some time to the air, it usually appears red and bloated. In many cases a watery froth, either white or blood-stained, will be noticed around the nostrils where the body has not been out of the water longer than a day.

A corrugated condition of the palms of the hands and of the soles of the feet, sometimes called "washerwomen's or cholera hand" may be noted, but is not indicative of submersion during life.

The pupils are commonly dilated.

Abrasions or excoriations of the fingers are commonly present, and gravel, sand, mud, etc., are often found under the finger-nails, and fragments of weeds clutched in the hands.

The penis is usually found contracted and retracted, and the scrotum shrunken and wrinkled.

Post-mortem rigidity is generally present and sets in so rapidly that not unfrequently the body remains stiffened in the last attitude of life.

Internal Appearances.   The brain in cases of drowning is sometimes congested, but not as . a rule, very hyperæmic.   The blood is generally fluid and of a dark color from want of aeration.   The right side of the heart, pulmonary arteries, and venous system are commonly congested with a dark fluid blood, although sometimes the right cavity of the heart is empty, and not unfrequently the two sides of the heart contain equal quantities of blood.

The larynx, trachea, and bronchi are commonly deeply congested, and the lungs also are usually more or less congested, and so distended as to completely fill the chest.   The presence of water in the pulmonary vesicles is strong evidence of submersion during life.   Dr. Ogston states, however, that in 48.7 per cent of cases by drowning, no water was found in the air-cells of the lungs.   Commonly, though not invariably, a froth, or rather a lather, formed of air, water, and mucus, tinged occasionally with blood, will be found in the air passages and around the lips and nostrils ; although this is usually present, its absence is not proof that the death did not result from drowning.

The stomach and alimentary canal are often much discolored.   The presence of water in the stomach, swallowed during efforts to breathe, is a fairly constant appearance ; its quantity and quality should be carefully noted, remembering that a little water in a water drinker's stomach is not significant.   If the quantity in the stomach is above half a pint, it may be regarded as having an important bearing upon the cause of death. Its quality may also afford important evidence.

The liver, spleen, and kidneys are usually gorged with blood ; the bladder rarely contains bloody urine.   The nose, lips, fingers, toes, genitals, etc., are sometimes found to have been gnawed by voracious fish.

The question as to how soon after death by drowning a body rises to the surface, depends upon a variety of circumstances, such as the season of the year, — bodies rising more rapidly in warm weather; the depth of the water, — bodies rising more easily in shallow than in deep water; the character of the dress ; weights or mechanical impediments to the rising of the body; the nature of the body, being fat or lean, and the specific gravity of the water, — bodies rising less rapidly in fresh than in salt water.   Dr. Tidy states that, as a rule, the body floats after from five to eight days, and that the popular notion of a body floating in three days is contrary to his experience, although it may in shallow water float as soon as twelve hours after submersion.

. M. Devergie's conclusions as to how long the body has been under water, formed on a large experience at the Paris morgue, are substantially as follows : —

During the first four or five days there is little change ; post-mortem rigidity may in some cases continue to the fourth day, especially if the water is cold.

The fourth or fifth day the skin of the palms, and particularly that of the ball of the thumb and the little finger, and the lateral surface of the fingers begin to whiten.   On the sixth or eighth days this extends to the soles of the feet. The skin of the face softens, and is of a more or less faded white than the rest of the body.

On the fifteenth day the face is slightly swollen and red; a green spot begins to form on the skin of the mid-sternum; the hands and feet are quite white, except the dorsum of the latter; the skin of the palms is wrinkled; the subcutaneous cellular tissue of the thorax is of a red color, and the upper part of the cortex of the brain is green.

In one month the face is reddish-brown, the eyelids and lips green and swollen, and the neck slightly green; a brown spot with green areola, about six inches in diameter, occupies the upper and middle part of the sternum. The skin is very wrinkled and the hair and nails still adherent. The scrotum and penis are much distended by gas, so that the latter is sometimes erect. The lungs are emphysematous with the gas of putrefaction.

At six or seven weeks the neck and thorax are very green, and the cuticle at the wrists begins to be detached.

At two months the body is covered with slime, which penetrates through the clothes; the face is enormously swollen and brown, the lips parted so as to expose the teeth. The skin on the middle of the abdomen, on the arms, forearms, thighs, and legs continues natural. From this time the skin with the nails attached begins to come off like a glove from the hands and feet; the hair falls off or can be easily detached by pulling; the veins are almost completely empty of blood and filled with gas. If, at the moment of death the right cavities of the heart were gorged with blood, the internal surface of the right ventricle will appear of a jet-black color.

At two and a half months the green color of the skin extends to the arms, forearms, and legs; the nails are quite detached. Some adipocere will be formed on the cheeks, chin, breasts, armpits and internal parts of the thighs; the abdomen is greatly distended by gas. The muscles are not yet much altered in color.

At three and a half months the scalp, eyelids, and nose are destroyed. The skin of the breast is generally of a greenish-brown color; the centre of the abdomen presents an opaline appearance, scattered over with small erosions caused by the water; larger erosions are found on other parts of the body; the hands and feet are bare of skin. There is a space between the lungs and the *pleura costalis* filled with reddish serum.

At four and a half months the face and scalp are so destroyed as to leave the skull bare, — the remains of the face, neck, and part of the thighs being converted into adipocere. As to the formation of this substance, see page 67, *ante.*

Small eminences indicating the commencement of calcareous incrustations are observed on the prominent parts. The brain presents traces of adipocere on its anterior parts.

The changes above described will proceed more rapidly in very hot weather or in putrid pools and ponds, but more slowly in cold water or in salt water, or where the body is closely invested by clothing. The general order of events, however, remains the same.

In determining the question whether drowning was the cause of the death, or whether the person was already

dead when thrown into the water, it should be remembered that no one post-mortem appearance can be regarded as conclusive; a correct conclusion can only be
formed by the consideration of all the facts revealed by
the autopsy. What these facts are has already been
stated. It should be remembered, however, that the
most characteristic signs of death by drowning are not
permanent, and that their continuance is shorter in summer than in winter; while the characteristic signs
may remain fifteen days in winter, they may disappear
within three days in summer. In a post-mortem after
a supposed drowning, all abnormal and diseased conditions, such as might of themselves have caused death,
should be noted, it being possible that the body has
been thrown into the water after death from natural
causes.

Very little water is required to drown; so long as
the face is covered it is sufficient.

**The determination of the question whether the drown-
ing was accidental, suicidal, or homicidal,** in the absence
of marks of injury, or even in the presence of marks
which may have been self-inflicted or caused by the
water or by objects in the water, is very difficult, if not
impossible. The presence, however, of certain marks of
violence, etc., such as a cord around the neck, stabs,
pistol wounds, etc., is suggestive of homicide. The
question whether the wounds were inflicted before death
and were the cause of the death may sometimes be determined by considerations stated in the chapter on
Wounds. Whether they are suicidal, accidental, or homicidal is frequently impossible to determine; and when
it can be determined, must depend upon the application

of the ordinary rules of evidence and not upon the medical expert.

Death by Hanging. By death by hanging is meant death caused by the partial or total suspension of the body by the neck by means of a ligature of some sort, the constricting force being the weight of the body itself.

In hanging, death may result from asphyxia, cerebral hyperæmia, a combination of asphyxia with apoplexy, syncope, or injury to the spinal cord and the pneumogastric nerves. According to Casper, out of eighty-five cases nine cases were caused by apoplexy, fourteen cases by asphyxia, and sixty-two cases by mixed conditions. Whether death resulted from one or the other of these causes will depend upon the tightness of the ligature, the position at which it crosses the neck, and whether or not force is employed, as well as the degree of force. Where the ligature is very tight, or a loose ligature crosses the neck above the os hyoides, asphyxia will predominate over coma ; where the ligature is loose, and the larynx is protected by the cord pressing against the os hyoides, then coma will predominate over asphyxia. A combination of asphyxia and coma is commonly the cause of death in cases of hanging by suicide, and where no violence has been exerted, the ligature both preventing the return of blood from the head,— thereby inducing congestion of the brain, — and preventing the entrance of air to the lungs, thereby causing apnœa.

Dr. Tidy states his conclusions thus : "(1) Given pressure both on the air-tubes and blood-vessels, pressure on the air-tubes being only partial, death will probably result from a combination of asphyxia and apoplexy, but from asphyxia primarily. (2) Given a pressure in such a position that the air-way is more or less protected, death may occur from apoplexy and will then be slow. (3) Given a complete pressure, so that the entrance of air into the lungs is entirely prevented, death will result from asphyxia, and will be rapid and possibly even instantaneous."

In cases where violence has been exerted, as in felonious and homicidal cases of hanging, when sufficient force is applied to break the transverse and other ligaments, or to fracture the odontoid process of the axis, death will be rapid, if not instantaneous. Under such circumstances the medulla oblongata can scarcely escape severe injury; and as this contains the centres of respiration and circulation, death will ordinarily be instantaneous.

In most cases of hanging unattended with violence, there are three distinct stages: first, a short stage of semi-unconsciousness, lasting from thirty seconds to three minutes; secondly, a stage of subjective death but of objective life, varying from ten minutes onwards; and lastly, a stage of general death, lasting until the occurrence of *rigor mortis.*

Without repeating those appearances common to death by apnœa or coma, **the post-mortem appearances** after death by hanging are as follows: —

The body may be found in *rigor mortis* in almost any position. The position of the head will vary according to the position of the knot or of the ligature; the usual position in suicide is for the head to be forcibly flexed forwards; the face is sometimes pale

but more often swollen and congested; the tongue is
usually enlarged and livid, either protruding or compressed between the teeth; the eyes are nearly always
staring and prominent and the pupils dilated; blood-
stained froth is sometimes found about the nose and
lips; the fists are often shut down so tightly that the
finger-nails penetrate the palms; the neck usually appears stretched, and will probably show the marks of
the ligature.   If the ligature is very soft and the body
cut down instantly after death there may be no mark
of the ligature; the mark, when it exists, is usually
oblique, following the line of the lower jaw; it is usually non-continuous where there is only one turn of the
cord, but where it has passed around the neck more than
once one mark may be circular and the other oblique.
The appearance and character of the mark of course
vary with the nature of the ligature and its method of
application, the vitality of the tissues, and the period
that has elapsed after death.   Its size and depth do not
necessarily correspond with the size of the ligature, although the narrower the ligature, and the longer the
suspension, the deeper as a rule will be the mark.
It is usually a well-defined groove or furrow, single or
double, regular or irregular, like its cause.   If the person
is young and the tissues healthy and the suspension brief,
the mark may be a slight depression without change of
color or at most a red blush; more frequently, however,
the bottom of the furrow appears white, and its edges
are usually slightly raised and red.   The most common
appearance, however, according to Dr. Tidy, presented
by the mark is that of a dry, hard, yellowish-brown,
parchment-like furrow; which condition was found by

Dr. Ogston in 32½ per cent of his cases. This condition, however, is not apparent till the body has remained suspended after death for several hours, and it does not prove suspension during life.

The state of the genital organs is often one of turgescence; the penis is often more or less erect, and there is often an emission of seminal or prostatic fluid; the urine and fæces are sometimes expelled; there is invariably a flow of saliva from the mouth, which being a vital act, may be regarded as evidence that the suspension took place during life.

The internal post-mortem appearances in cases of hanging other than those described under the head of asphyxia and coma, are the fracture or dislocation of the cartilages of the larynx and, rarely, the fracture of the os hyoides; there may also be a dislocation of the cervical vertebrae, accompanied with rupture of the ligaments and fracture of the odontoid process; the inner and middle coats of the carotid artery may be ruptured; there may also be lesions of the skin or of the deeper-lying soft parts; the larynx and trachea are usually deeply congested; the condition of the lungs and heart will vary accordingly as death resulted from syncope, asphyxia, etc.; the stomach is often so congested as to resemble the effects of irritant poisoning; the cerebral vessels are rarely much congested.

As to whether hanging was the cause of death or not, it must be remembered that the mark of a cord is not conclusive proof of death by hanging; nor is its absence conclusive evidence that death did not result from hanging. The fact that the body is found in such a position that the feet can touch the ground is not evidence that the death did not result from hanging. Suicide by hanging is consistent with almost any posture of the body; persons so dying have even hanged themselves lying at full length on a bed. On the whole, the medical jurist can seldom give a certain answer to the question whether death was caused by hanging.

The determination of the question whether the hanging was accidental, suicidal, or homicidal must usually depend upon the circumstances of the case, and the general rules of evidence, with which the medical jurist has no concern. It may be remarked, however, that the probabilities are in favor of suicide; it must be remembered also that hanging may occur by accident.

Death by Strangulation. — By death by strangulation is meant death resulting from pressure on the neck applied otherwise than by the weight of the body itself, — as by the fingers and thumb, knee, ligatures, etc. Death by strangulation is generally due to apnœa, and hence the post-mortem appearances of asphyxia are usually more marked than in death by hanging. The face and extremities are usually livid and swollen; there are minute ecchymosed spots on the skin of the face, neck, chest, and conjunctivæ; blood occasionally issues from the mouth, nostrils, ears, and eyes; the eyes are congested and prominent, usually wide open with dilated pupils; the tongue is frequently swollen, protruded, dark-colored, and sometimes bitten; the hands are commonly clenched. The marks on the neck will vary according to the nature of the force employed. Where the cause is manual pressure, marks of the thumb and one or more fingers, together with scratches caused by the nails, are usually found on the front of the neck. Where strangulation is caused by a ligature, a mark usually more entirely encircles the neck than in hanging, and is generally lower and less oblique. Where the pressure is caused by a hard body wrapped in a handkerchief, etc., there may be bruises of considerable size at one spot on the neck. Generally effused blood will be found in the subcutaneous areolar tissue and muscles under the mark. The inner and middle coats of the carotid arteries may be ruptured; the lining membranes of the larynx and trachea are always more or less congested; extreme injury to the neck, though it sometimes occurs, is not common. In homicidal throttling extensive lesions of the larynx usually occur. The brain is sometimes congested but usually normal; the lungs are sometimes congested and sometimes normal; sometimes patches of emphysema due to the rupture of the superficial air-cells, either singly or in groups, are found in the lungs; the heart is sometimes empty and sometimes full of dark fluid blood; the genitals of both male and female are sometimes congested; involuntary discharges of urine, fæces and seminal fluid may or may not occur; the stomach is often congested, sometimes normal; the blood is usually very dark and very fluid.

It is very difficult to determine whether death was actually caused in any instance by strangulation. A medical jurist would not be justified in pronouncing death to be the result of strangulation by anything short of distinct external marks.

Whether a given case of strangulation was suicidal or homicidal is a difficult question, and must be determined by the ordinary rules of evidence.

Death by Suffocation. — By suffocation is meant the

exclusion of fresh air from the lungs by means other than external pressure on the trachea. Within this definition are included both drowning and smothering. Drowning has already been considered ; and in this connection death by smothering alone will be considered.

Suffocation may be variously caused, — by direct pressure on the thorax ; by covering the head with bedclothes, shawls, etc. ; in case of infants, by overlying ; or by anything which will prevent the entrance of air into the lungs. The air passages may be closed internally by foreign bodies either in the air passages or in the œsophagus. Suffocation has also been caused from obstructions caused by the tongue, epiglottis, and velum palati, during the administration of anæsthetics while the patient is lying down. It has also been caused by blood from the nose and from wounds of the mouth or throat ; by scalds or other irritations of the epiglottis ; tumors pressing on the throat ; the bursting of an abscess in the pharynx or tonsils ; and in various other methods too numerous to mention.

In cases of complete suffocation, experiments render it probable that death will occur on an average in from two to five minutes.

The post-mortem appearances in death from suffocation are principally those of asphyxia already described, although it is said these appearances are not so well-marked as might be expected. If external violence has been used there may be flattening of the nose and lips ; ecchymosed scratches upon the throat may be found. Patches of lividity, and of dotted or punctiform ecchymoses will usually be found on the skin and conjunctivæ ; the lips and extremities are usually livid ; the face

may be pale or violet, but is often placid; the eyes are
usually congested; mucus and sometimes blood-stained
froth is found about the mouth and nose; the blood is
usually dark-colored and very liquid; the brain and the
vessels of the pia mater are generally congested; frequently the right side of the heart is more or less full of
blood, but occasionally the heart has been found empty;
the trachea is usually of a bright-red color, and often
contains bloody froth; the œsophagus and trachea frequently exhibit evidence of injury; the lungs are sometimes very congested, at other times normal.

Tardieu and others lay a special stress on the existence of punctiform subpleural ecchymoses, which are
usually found at the root, base, and lower margin of the
lungs.   That these punctiform ecchymoses are not infallible signs of death from smothering, seems to have
been proved by Dr. Ogston, for they are sometimes absent in cases of death by suffocation, and they may be
present when death is due to some other cause.

The kidneys are generally congested.   In a post-mortem examination in a case of suspected smothering, careful examination should be made of the air passages for
foreign bodies.

As there is no absolutely certain anatomical appearance characteristic of death by strangulation, the question whether death is so caused does not admit of a
positive answer from post-mortem appearances alone;
but must be determined from these in connection with
the other evidence of the case.

The question whether the death was accidental, suicidal, or homicidal is one to be determined by the legal
rather than the medical jurist.

CHAPTER VI.

The average temperature of the body in health is 98.6° F., or 37° C. in the mouth and axilla; or from .9° to 1.3° F. higher in the vagina or rectum. The temperature may range 1.8° F. above or below the above average consistently with health. In disease or after accidents, the temperature may rise, or fall below the normal. In certain diseases it may range as high even as 115° F., and there are recorded instances of the temperature falling as low as 75° F.; but such high or low temperatures are very exceptional. It may be stated as a general rule that if the temperature of a warm-blooded animal be raised for any length of time by any means to the extent of from 11 to 13 degrees F., death is certain; and that if the normal temperature is for any length of time depressed from 18 to 27 degrees F., death is equally certain. The very young and the very old have limited powers of heat production, and cannot therefore well endure extreme cold. Young adults bear cold the best, and young males better than females of the same age. With proper precautions, however, extreme heat and extreme cold may be tolerated for some time. Death from cold is usually accidental, although there are recorded instances of the application of cold for the purpose of homicide, especially in the case of

new-born infants. A new-born child, if left unclothed or
exposed to the air in a cold room, will soon die. Insane
persons are said to be more susceptible to the effects of
cold than sane persons, although frequently insensible
so far as their feelings are concerned. In the cases of
wounded persons, exposure to cold will also undoubtedly
increase their danger.

As to the post-mortem appearances in death from cold,
rigor mortis generally sets in slowly and lasts for a long
time. Dr. Ogston, from the inspection of sixteen bodies
after death from cold, comes to the conclusion that where
all the hereinafter described appearances are encountered in the same case, in the absence of any other obvious cause of death, they point with high probability
to the death's having been caused by cold. In adults
these peculiar appearances are : —

First, an arterial hue of the blood generally, except
when viewed in mass within the heart. The presence
of this coloration was not noted in two cases.

Second, an unusual accumulation of blood on both
sides of the heart, and in the larger blood-vessels, arterial
and venous, of the chest.

Third, pallor of the general surface of the body, and
anæmia of the viscera most largely supplied with blood.
The only exceptions to this were moderate congestions of
the brain in three cases, and of the liver in seven.

Fourth, irregular and diffused dusky-red patches on
limited portions of the exterior of the body encountered
in non-dependent parts ; these patches contrast forcibly
with the pallor of the skin and general surface. "The
above appearances," says Ogston, " were not, however, so
universally met with in the children as in the adults.

The arterial hue of the blood was absent in one; the anæmia of the larger viscera in all but one instance. The pallor of the surface was present, nevertheless, in all but one of the children, and the dusky-red patches on the whole of them." He states in conclusion, however, that "the subject of death by cold, so far as it can be ascertained by the inspection of the dead body, requires further elucidation than has yet been bestowed upon it."

. Certain parts of the body may be found frozen, but this is usually a post-mortem phenomenon; putrefaction does not occur at a freezing temperature.

Death by Heat. — Respecting the immediate cause of death by heat, "our ignorance" says Dr. Ogston, "is still great." During the heated season death from sun-stroke, insolation, or thermic fever is not uncommon. Something more than mere heat is probably necessary to induce thermic fever; for there is a recorded instance where a man in good health has remained in the hot room of a Turkish bath continuously for four days without injury to his health, the day temperature averaging 140° F., and the night temperature 125.°

There are three classes of cases included under the general term "sun-stroke." The first, which is said to be very rare, is acute meningitis or phrenitis. In the second, there is heat exhaustion with collapse, a rapid, feeble pulse, cold moist skin and a tendency to syncope. In the third class there is thermic fever; and of this fever the name "sun-stroke" is considered by Dr. Wood a misnomer, — it being caused by the action of external heat independently of its source. The phenomena of sun-stroke and heat apoplexy are said to depend on the action of the superheated blood upon the nerve centres and large internal organs. But for the further consideration of this subject the student must be referred to technical works on the practice of medicine.

The post-mortem appearances are, as a rule, negative; that is, there are no constant lesions, and in some cases nothing abnormal is found.

As to the so-called cases of **spontaneous combustion**, while it is possible that under certain circumstances, as for instance, in hard drinkers, the body may be preternaturally combustible, the alleged cases of spontaneous human combustion do not seem to be entitled to much consideration.

Death by Lightning or Electricity. — In cases of death by lightning, death may be caused by shock, by the severity of resulting burns and wounds, or by the disorganization of tissues and rupture of structures necessary to life.

Where the death has been instantaneous, characteristic appearances are not invariably met with in the body. Burns and blisters are common effects; wounds and a bruised condition of the parts are not uncommon; livid streaks and ecchymosed spots are frequent; frequently the marks assume a peculiar arborescent appearance depending upon the course of the veins, or the disposition of metallic bodies about the person, which marks are said to be very indicative of death by lightning. Various nervous and other symptoms often follow a severe shock by lightning which does not cause death.

The indications of this form of death are strengthened by evidence in the vicinity of the dead body of the effect of the electric current on trees, buildings, etc., or upon domestic animals in the vicinity.

Death by Starvation. — There are two sorts of starvation, chronic and acute. By the former is meant the withholding of food either sufficient in quantity or proper in quality to support life. It may also result from mal-assimilation of food in disease, nausea in pregnancy, and from other causes.

By acute starvation is meant the deprivation of all food from a person previously well fed. In chronic starvation hunger is not a marked symptom; in acute starvation intense hunger appears to be a symptom of comparatively short duration.

The general symptoms of starvation appear to be those of great nervous depression; the pulse is usually slow and soft; the features collapsed; the voice hollow and the breath offensive. There is usually a marked debil- · ity, with great languor, listlessness, irritability, and despondency. The skin is usually harsh and dry, and frequently, especially in chronic cases, becomes covered with a brownish-colored, filthy-looking coating. There is great emaciation; pain and irritation in the stomach are usually troublesome. As a rule, the bowels are very costive and the body generally feels cold. The mucous membranes of the outlets of the body are frequently red and inflamed; the temperature is generally lowered; there is great loss of weight, — amounting, it is said, in fatal cases, to an average of forty per cent of the entire weight.

As to the time of death in starvation, much depends upon the previous condition of the body, access to fresh water, existence of warmth, age, etc. Old people seem to bear want of nourishment best; middle-aged next; those just arrived at puberty are less able to endure it, and children are the least able. The time of the recorded cases of death from starvation, without taking into account access to water, varies from the seventh to the sixtieth day. The average period of death in case of complete deprivation of food would seem to be from seven to ten days. The longest authentic period of total deprivation of food in human beings, according to Woodman and Tidy, is about six weeks. In the case of Griscom, reported by Dr. Lester Curtis, of Chicago, the fast extended forty-five days without fatal results. The cases of deprivation of food longer than this, and most of those much shorter, may well be regarded with suspicion.

Our knowledge of **the post-mortem appearances in cases of starvation** is principally derived from chronic cases. In such cases the body is always greatly emaciated, and there is an almost entire absence of fat; sometimes, however, one part is much more emaciated than another. The condition of the skin has already been noted; the muscles are commonly soft, pale, and wasted; the heart is generally not quite empty; it is usually, however, more or less contracted, and sometimes soft and bloodless. The œsophagus is usually small and contracted. The condition of the stomach has been variously recorded as natural, as small and contracted, as corrugated, as loose and flabby. It frequently contains a little dark, gelatinous fluid. The small intestines are generally contracted both in length and calibre, and are commonly thin and transparent; they are sometimes empty, sometimes contain a little dirty mucus, and are sometimes distended with gas. The large intestines are usually transparent, and frequently contain hard, fæcal matter ; the omentum is usually transparent; the liver is usually healthy but contracted ; the gall-bladder is usually full; the pancreas is invariably atrophied, and the bladder invariably empty ; the tissues of the uterus are often soft and relaxed.

A contracted state of the stomach and bladder, and a shrunken and transparent condition of the intestines and omentum, with a more or less atrophied but otherwise healthy condition of the viscera, appear to be the prominent post-mortem appearances after death from starvation.

CHAPTER VII.

THE general opinion of the profession is generally believed, at least up to a very recent period, to be the same as that expressed by Dr. Emmett, who in his work on Gynecology has said, "Pregnancy may be suspected, but we have no sure and at the same time lawful means of proving its existence before quickening, or until the enlarged uterus has risen from the pelvis into the abdominal cavity; when this has occurred, if pregnancy exists the beating of the fœtal heart may be detected, or the motion of the child may be felt, and either of these must be accepted as conclusive evidence."

While this statement is probably correct, still, pregnancy can hardly exist without furnishing evidence of its existence; and although perhaps its existence cannot be demonstrated conclusively before quickening, it may under many circumstances be made highly probable.

In determining whether a living woman is, or is not pregnant, it should be emphasized at the outset that the medical jurist should not rely upon the statements of the woman herself. The signs which present themselves prior to the end of the fifth month or the begin-

ning of the sixth are not to be regarded as entirely
conclusive; among these may be mentioned —

1. The cessation of the menses.

2. Various sympathetic disorders, such as morning
sickness, vomiting, loss of appetite, salivation, headache,
and toothache. Morning sickness commonly commences from the second to the sixth week after conception, and generally ends about the fourth month.

3. Changes in the mammæ. These changes usually
begin with pregnancy and become clearly perceptible
at the end of six weeks or two months. The breasts
grow larger, firmer, and more knotty. At the end of
six or seven weeks an areola or decided darkening
(this only applies to a first pregnancy, as it is apt to
retain its color and breadth after women have borne
one or more children) will be noticed around the
nipple, varying in diameter from one half to two or
three inches. Upon this dark ground are to be noted
from twelve to twenty follicles or tubercles of a lighter
color. The superficial veins also become prominent as
pregnancy advances, and sometimes circular streaks will
be seen by slightly stretching the skin. The nipples
become more prominent and swollen, and now begin to
secrete a milk-like fluid which may be squeezed therefrom. The skin around the nipples is in most cases
soft and moist.

4. Abdominal symptoms. During the first two months
of pregnancy there is little change in the size of the
abdomen; if anything it is a little flatter. About the
third month, however, the abdomen begins to increase
in size; the skin is gradually stretched and the navel
obliterated; and from this time the abdomen steadily

7

increases in size till nearly the end of gestation. At the end of the eighth month, when the uterus has reached the ensiform cartilage, the navel becomes very prominent.

Dark pigment-cells are deposited, especially along the mesial line of the abdomen, extending from the pubes to the umbilicus or the sternum. They are most marked in a dark-complexioned woman.

5. Changes in the vagina and uterus. The vagina in pregnancy is generally somewhat relaxed, its mucous membrane congested, giving it a violet tinge; the inner surface of the vulva also presents the same appearance. This condition generally continues until the fourth or fifth month. In this connection mention should be made of the state of the urine during pregnancy. The deposit known as "kiestine" was once thought to be diagnostic of pregnancy, but at the present time it is regarded as a sign of little value.

The state of the *os* and *cervix uteri* deserves particular attention. Soon after impregnation, on digital examination *per vaginam*, the *cervix* will be felt to be fuller, rounder, and more spongy or elastic than in its unimpregnated state, while its orifice will have lost its transverse shape and well-defined edge and become rounder and thicker. Gradually, as the period of pregnancy advances, it not only becomes less prominent, but from and after the sixth month more and more flattened, till about the time of delivery, when it can no longer be felt. For the first three months it is low down in the vagina; a little earlier than the fifth month it rises with the ascent of the uterus into the abdomen, and from this period it recedes farther from

the external parts, and ceases to be distinguishable as
a projecting body. The absence of these changes in the
os or *cervix uteri* is of more value than their presence.

The changes in the body of the uterus are important.
During the first three months the degree of its development is ascertained with difficulty; by the end of the
fourth month the fundus of the uterus may be felt,
especially in a thin person, above the anterior wall of
the pelvis. During the fifth month it has usually risen
between the pubes and the umbilicus. During the
sixth month the fundus rises as high as the umbilicus.
The seventh month it is on a level with a point midway
between the umbilicus and lower end of the sternum;
and at the end of the eighth month it has reached its
highest point, the level of the ensiform cartilage, from
which it settles to a somewhat lower position before
delivery.

A new sign has recently been brought to the notice
of the profession by Dr. Hegar of Freiburg, which seems
to be of considerable promise and has recently attracted
much attention. This sign is said to consist of an
unusual resilience, compressibility, softness, bogginess,
yielding and thinning of the lower uterine segment, —
that is, the section immediately above the insertion of
the sacro-uterine ligaments. The remainder of the body
of the uterus is often firm and hard, and its shape is
stated to be more fan-like, or like that of a balloon, than
the usual pear shape. This enlargement is especially
marked antero-posteriorly; the change is most apparent
at the middle portion of the lower segment in the mesial
line, the sides of the organ being much firmer and more
resistant.

During the fourth month and thereafter, a low murmuring or cooing sound, like that made by gently blowing over the mouth of a wide bottle, but with no impulse, may sometimes be made out; this is called the "uterine or placental souffle;" this, however, is not a certain sign of pregnancy, as it may be caused by a tumor pressing upon the aorta or the iliac vessels.

6. Evidence afforded by the condition of the fœtus. This subject may be considered under three heads: first, its passive movements; secondly, indications of its vitality afforded by the stethoscope; and thirdly, active movements of the fœtus, usually called quickening.

a. The passive movements of the fœtus may be demonstrated by what is known as "balottement;" that is to say, an impulse given to the fœtus by the finger applied to the *os uteri* through the vagina while the woman is in a standing position, or lying with the trunk in a semi-recumbent position, the fœtus being raised through the *liquor amnii* by the finger giving the impulse, to return again to its original position at the lowest part of the uterus. It is seldom satisfactory earlier than about the end of the fourth, or later than about the end of the sixth month of utero gestation. This sign is a sign of considerable value, but probably not conclusive. The following signs are, however, conclusive.

b. By the use of the **stethoscope** applied to the uterine region of the abdomen, the pulsations of the fœtal heart may frequently be heard, although they can, according to Dr. Tidy, seldom be satisfactorily made out earlier than the end of the fifth month. According to Dr. Lusk, they may generally be made out by auscultation by the eighteenth to the twentieth week; under favorable circumstances they have been detected as early as the fifteenth to the sixteenth week. They resemble the ticking of a watch under a pillow, and vary in frequency from 120 to 160 beats per minute.

c. The active movements of the fœtus afford further conclusive proof of pregnancy; these active movements, called quickening, are, according to Dr. Tidy, experienced usually by the pregnant female between the sixteenth and the twenty-fourth week, but may occur as early as the twelfth. According to Dr. Lusk, modern investigations place the time at which the fœtus first begins to employ its muscles at about the tenth week. It is, however, according to him, somewhat rare for these movements to excite the attention of the mother before the sixteenth to the eighteenth week, though experienced matrons may recognize them at an earlier period. As a subjective symptom quickening is of little value, but as an objective symptom where care is taken not to confound it with the contractions of the uterus itself or with the abdominal muscles, it is conclusive. This fœtal impulse may, according to Dr. Tidy, often be felt externally about the third or fourth month. According to Lusk, active fœtal movements, such as may be recognized by the medical expert, seldom assume much distinctness before the sixth month. When recognized distinctly they are conclusive.

The signs of pregnancy discoverable by a post-mortem examination are, aside from the subjective symptoms, chiefly those which would be observed in the living; besides these, however, the presence of the ovum or the existence of a distinct fœtus with its placenta and

membranes would be demonstrated upon a post-mortem
if the woman were actually pregnant.

A true *corpus luteum* should be present in one or both
ovaries. By true *corpus luteum*, or the *corpus luteum* of
pregnancy, is meant the cicatrice formed after the discharge of the impregnated ovum; by a false *corpus luteum*
or menstrual *corpus luteum*, is meant that which is formed
after the discharge of the unimpregnated ovum at each
menstrual period. This subject has in the past afforded
a fruitful field of controversy. Without entering into
the full consideration of the subject (for which the
reader is referred to standard works on obstetrics and
physiology) it may be said that the subject has lost
much of the importance formerly attributed to it, for
the reason that pregnancy may exist without the formation of any true *corpus luteum*, and that *corpora lutea*
may be found where there has been no pregnancy whatever, and that even in aged women long past the child-
bearing period.

The question sometimes arises for decision by the
medical jurist, whether a woman has or has not been
pregnant at some period in the past, and may be particularly important in certain cases of disputed identity
after death. The presence of an intact hymen is of
course very strong evidence that no mature or nearly
mature child has been born in the natural way. Cæsarean section would leave a long scar. If the breasts
and genital organs preserve their elasticity and virginal
character, the evidence is strong although not conclusive against previous pregnancy. The reader is
referred in this connection to the signs of pregnancy,
ante. As a general rule, no absolutely certain opinion

can be given at the autopsy whether a woman has or
has not borne children; but as a question of probability some reliance may be placed on the internal
appearance of the uterine walls, more especially as
respects their convexity. In the multiparous uterus,
the anterior and posterior surfaces of the body of the
uterus are more rounded than in the virgin uterus. The
fundus instead of being flat is convex, so that there is
considerable protuberance above a line drawn-from one
Fallopian tube to the other; the vaginal portion of the
neck is altered, being usually larger and more prominent
in the vagina. The *os uteri*, instead of presenting a
transverse fissure or smooth round aperture, is more
open and puckered; the depression is more evident, and
the orifice considerably larger. The uterus after gestation rarely returns to the size of the nulliparous organ;
its diameters are all increased; the cavity of the body
of the multiparous uterus is also considerably enlarged;
the *os internum* is less distinct and the canal of the
cervix shorter; the penniform rugæ are considerably
obliterated; the cavity of the body is less distinctly triangular in shape, — the angles into which the Fallopian
tubes enter being less marked. Again, if the posterior
commissure or fourchette be intact it is morally certain
that the woman has not given birth to a child at full
term. The posterior commissure is, however, rarely
affected by sexual intercourse, even in the case of prostitutes; although it may be destroyed by extreme violence short of the birth of a child, its ruptured condition
is strong evidence of the woman's having borne a child,
while its intact condition is strong evidence of non-
delivery, or at least of non-delivery at full term.

As to the time which must elapse after delivery before a woman can again become pregnant, it is to be remarked that the time for the restoration of the genital organs to their normal condition is very different in different women; and the time which must intervene between the delivery of one child and the conception of a second has been variously stated at from fourteen days to one month.

· The subject of the duration of pregnancy, superfœtation and other kindred subjects, will be considered in the chapter on LEGITIMACY AND PATERNITY.

In concluding this chapter, it should be always remembered that a medical man has no right to examine a woman in a case of suspected pregnancy without her full consent, unless under the order of a court of competent jurisdiction.

CHAPTER VIII.

The signs of recent delivery in the living may become a matter of inquiry in cases of suspected concealment of birth and child-murder as well as in some other cases. The degree of certainty in such inquiries depends much upon the time that has elapsed since the birth of the child. The evidence to be derived from an inspection of the child will be considered in another connection. If the examination of the alleged mother be conducted within a week, most of the following symptoms will be present; but if the examination be delayed much beyond a week or ten days the evidence of recent delivery will be inconclusive if not negative. The signs of recent delivery at or near full term, are as follows : —

A peculiar pallid expression of countenance; eyes sunken, with a dark areola under and around them; a peculiar odor of the body is also present; the skin is usually soft, moist, and relaxed; the pulse will be a little quickened and more than usually soft and compressible.

The time occupied by the following signs is divided by Dr. Ogston into three periods : —

The first period which is of an average duration of forty-eight hours, embraces the time from the moment of delivery to the approach of the milk fever.[1] If the

---

[1] According to Lusk, since the general introduction of the thermometer into practice, and the better understanding of the causes of

woman be examined during this period the vulva will be found gaping; the labia and nymphæ torn and swollen, and the fourchette, if the delivery is the first one, lacerated; the vagina will be found soft and dilated and more or less bathed in mucus; the mouth of the uterus very much dilated and soft. The uterus itself may be felt to be quite bulky by the hand placed above the pubes; the abdomen itself will be relaxed and in folds; the brown line noticed as a sign of pregnancy may also be observed, and minute clefts, at first pink, afterwards white like ordinary scars, termed "lineæ albicantes," may be seen crossing each other in all directions.

At the end of some hours after delivery the lochial discharge commences, consisting at first of pure blood without special odor, and becoming toward the end of the second day, pale and watery. About the third or fourth day after delivery it is almost entirely suppressed under the influence of the milk fever.

The second period, which continues usually from thirty-six to forty-eight hours, includes the duration of the milk fever and the swelling of the breasts. The so-called milk fever usually commences on the third day

febrile temperature in the puerperal state, the existence of a distinct milk fever referable to functional disturbances in the breasts during the period in question has been found to be an exceptional occurrence. Under normal conditions the temperatures of the third day do not rise above 100.5° F. With this sub-febrile increase there is, however, often considerable general disturbance, indicated by slight chilly sensations, headache, anorexia, and a quickened pulse, — which, however, disappear in the course of twenty-four hours, — with profuse perspiration and an abundant secretion of milk. As the term milk fever is, however, in common use, and there is physiologically at this time a temperature above that of health, it will be convenient in use to retain the term.

after delivery, but sometimes it begins on the first or second day, or even as late as the fourth or fifth; it is best marked in women who do not nurse their children; it is preceded by headache, heat and dryness of the skin; the pulse is at first small and hard, and then becomes fuller, and in a few hours the breasts swell; moisture on the surface and an abundance of acrid sweat succeeds. In from six to twenty-four hours the fever abates and a watery milk flows from the nipples, lessening the swelling and the tenderness of the breasts, which, however, disappear slowly.

The third period of from four to five days is indicated by the characteristic discharge of the lochia, which, about the fourth or fifth day, as the milk fever subsides, reappears as a more or less consistent yellowish-white fluid, with a characteristic odor distinguishing it from all other genital discharges. It quickly becomes seropurulent, and may thus continue from fifteeen days to three months, or even to the return of the menses; this fully insures the discharge of the fluids contained in the uterine walls, and the contraction of the uterus, which in five or six weeks has nearly resumed its former diminished volume in the unimpregnated state; for it never becomes quite so small as before. About this time, if the woman has not nursed her child, the menses usually reappear. The marks of contusion and distention about the vulva usually entirely disappear within two or three days after delivery, and it is chiefly during the first and second of the above periods, which are only approximative, that the fact of recent delivery can be properly verified. The most characteristic sign during the first period is the sanguineous discharge with

the odor of the *liquor amnii.* The milk fever in the
second stage may be very slight, and does not always
occur. The lochia proper which characterizes the third
period is a valuable sign, more conclusive when it continues for some time ; but when long continued it may
be confounded with the leucorrhœal discharge, into
which it sometimes changes. . . .

It should be remembered that the proof of recent
delivery at full term in the living female, can only
safely be predicted upon the presence of all or nearly
all of the signs above enumerated; and that after the
eighth or tenth day, these signs in general cease to be
distinguishable. According to Dr. Lusk, however, an
approximate estimate of the date of confinement may
be made during the first two weeks after delivery.

The signs of recent delivery before full term are the
same in kind, though less in degree than those at full
term, and are therefore much more liable to mislead.
In premature labor and in abortion, the signs of delivery, at whatever time investigated, will be found
indistinct in proportion to the immaturity of the ovum.
In premature labor in the later months, the signs of
delivery will be nearly as distinct as at full term ;
but after abortion at an early period, so little change
is made in the condition of the uterus and other parts,
and the woman may exhibit otherwise so few signs of
pregnancy even when examined a day or two after the
occurrence, that it may be found impossible to form
any definite opinion, unless the structure of the ovum
can be made out from the substance expelled from the
uterus.

It ought to be remembered, however, that the birth

of a child in the case of some women causes so slight
a disturbance as hardly to interrupt them in the pursuit
of their ordinary avocations, and that we are liable to
underrate the strength and endurance of a recently
delivered female. .

The question sometimes arises whether it is possible
for a woman to be delivered in a state of unconsciousness. Although such cases are very rare, there can be
no question that it is possible and has sometimes
occurred. There is one recorded instance of delivery
while in an induced hypnotic state.

As to the evidence of delivery in the remote past, it
is to be observed that it seldom admits of being clearly
proved. The absence of the fourchette, the perineal
cicatrice, the irregularity of the cervix uteri and the
milk in the breasts, are none of them conclusive ; milk
has been found in the breasts of virgins, old women,
and even of men.

**As to the signs of delivery discoverable upon post-
mortem**, besides those already mentioned, it may be
stated that shortly after delivery the articulations of
the pelvis are found to be more movable than usual ;
the shape of the uterus is more globular than in the
virginal state ; its walls are thicker, more spongy and
vascular, and its dimensions larger. Remains of the
decidua may be found lining the interior of the uterus
except where the placenta was attached, at which place
the uterine surface will either be raw and bloody, with
openings into the uterine sinuses, or will be covered
with mammillæ from albuminous deposits; and the
Fallopian tubes and openings will appear swollen and
vascular. These appearances gradually fade and noth-

ing satisfactory can be learned after the tenth day, and even then the examination will sometimes be fruitless.

With respect to the corpus luteum nothing need be added to what has already been said.

CHAPTER IX.

THIS topic regards primarily the infant, and the mother only secondarily. It is often a matter forensically of great importance to determine whether a child has or has not been born alive; also at times, whether a child at birth is mature or immature. The subjects of Infanticide and Abortion will be considered in another chapter.

The meaning of the term "live birth" at the common law is, " that there must be a manifestation of some certain sign or signs of life by the child after it is completely born ; " by which is meant that the child must be completely external to its mother. It is not, however, necessary to this definition that the cord should have been divided or the placenta separated. Such being the definition of live birth it follows that the time of a child's birth is the moment of its complete expulsion from the body of its mother.

The evidence of life derived from an examination of a child immediately after its birth, may consist —

1. Of muscular twitchings. Mere muscular movement independent of breathing, while very unlikely to be mechanical or independent of vital power, can hardly be regarded as sufficient evidence of live birth in a criminal inquiry; certainly not unless observed by an

experienced person accustomed to careful observation of
vital phenomena.

2. Respiration, by which we mean breathing as
evidenced by chest movements, is a certain sign of
life. The contrary, however, is not true; a child may
be alive without respiration.

3. Crying after complete delivery is conclusive proof
of live birth, but a child may be alive without crying.
It is also well settled that a child may in some rare
cases cry during the time of its expulsion from the
mother (*vagitus uterinus vel vaginalis*), and yet be completely dead when legally born; and also that a child
may be born alive and not cry for some time thereafter.

4. Pulsation in the cord, being caused by the umbilical or hypogastric arteries and due to the action of
the child's heart, is a clear sign of life.

5. The beating of the child's heart after complete
birth is the crucial test of live birth, and is sufficient
to establish this fact in the absence of all the other
signs of live birth.

The evidence of still birth may consist in the absence
of the foregoing signs of live birth.

The fact of the child's being dead when born will of
course be conclusively established if the body is found
in a state of putrefaction at the time of birth, — such
putrefaction being the result of intra-uterine maceration in the *liquor amnii*. The appearance of the body
so born dead is thus described by Casper: "It is impossible to mistake the appearance of a child born
putrid. The swollen cutis; the vesicular elevation of
the cuticle or its complete peeling off; the grayish-

green coloration of the body; the putrid navel-string; the well known stench, etc., do not constitute the diagnosis, since every child even when born alive, undergoes these putrefactive changes in their turn at the proper time after its death. On the contrary, most of these characteristics are not exhibited by a child born putrid. . . . In the first place a child born putrid is remarkable for its penetrating stench which cannot be concealed by a thin coffin or chest, etc.; and which though so repulsive and indestructible, is not yet the usual well known odor of putrefying bodies, but has something sweetish, stale and indescribable about it, which makes it all the more unendurable. . . . A child born putrid has not a shade of green upon its skin but is more or less of a coppery red, here and there of a pure flesh color; peeling of the cuticle is never absent, but close to recent patches of this character older ones are found upon the body, the bases of which are already dark and hardened. The excoriated patches are moist, greasy, and continually exude a stinking sero-sanguinolent fluid, which soaks through all the coverings of the body. The general form of such bodies is as remarkable as their color. While every highly putrefied corpse preserves for long the roundness of the contour of the body, though its form is disfigured and distorted by intumescence, it must strike every one, when a child born putrid is placed before him, how great a tendency is displayed by it to flatten out and, as it were, to fall to pieces. The thorax and abdomen lose their roundness; their contour forms an ellipse, from the soft parts sinking outwards towards both sides. The head itself (the bones of which are loose and movable, as in every

8

child's body) becomes flattened, and the face thereby repulsively disfigured, as the nose is flattened and the cheeks fall to opposite sides."

The fœtus may remain in the uterus months in this condition, and sometimes in twin pregnancies, by compression, one of the infants is caused to become flattened and atrophied so as exactly to resemble a little gingerbread figure.

The evidence of live or still birth afforded by a post-mortem examination of the body of a child, will be considered in the chapter on INFANTICIDE.

The subject of the maturity or immaturity of a new-born child, will be considered in the chapter on LEGITIMACY AND PATERNITY.

CHAPTER X.

WITHOUT undertaking a legal **definition** of the term abortion, which doubtless will be found to differ in different States, and without insisting upon the distinction which the accoucheur makes between abortion and premature labor, the term may in this connection be considered as applied to labor brought on at any stage of pregnancy, and may be considered under three heads.

1. Natural abortion, which includes miscarriage, or the expulsion of the ovum or of a non-viable fœtus; and premature labor, or the expulsion of the child after it is viable.

2. Artificial abortion, or the induction of premature labor for the purpose of saving the life of the mother and, if possible, of the child.

3. Criminal abortion.

The full consideration of natural and artificial abortion belongs to technical treatises upon obstetrics. It may be observed, however, that natural abortion may be caused by various accidental or pathological conditions of either the mother or the child, or may be caused partly by both. It is often very difficult to detect the cause, and it is beyond the scope of this work to enter into the further consideration of this subject.

As to artificial abortion, there can be no doubt that, after due consultation, with the object of saving the life of the mother or child, or both, operations for the purpose of inducing abortion or premature labor are entirely justifiable; indeed, their legality is sometimes recognized by statute. The cases where the induction of abortion or premature labor are justifiable are certain cases of pelvic deformity; some cases of obstinate vomiting during pregnancy; where pregnancy is complicated with insanity; and various other pathological conditions.

The term abortion is popularly understood as applying to the expulsion of the contents of the womb before the sixth month of gestation; but it is believed that the statutes upon the subject generally make no such distinction.

The production of abortion is generally made a statutory felony, and hence punishable by imprisonment for a greater or less time in the penitentiary. It should also be remembered that if the death of the mother is caused, as is not unfrequently the case, owing to the usual ignorance of the operators or the dangerous nature of the drugs employed, the crime is murder, although the accused had the full consent of the woman to the operation, and did not intend to cause death.

A full discussion of the legal questions involved belongs more properly to professed treatises upon the criminal law. In this connection we will consider only the means of producing abortion, and the evidence of abortion afforded by an examination of the female during life, or after death, or of the substances expelled from the womb, instruments or drugs used, etc.

The means used to produce criminal abortion are various. They usually consist in the administration of ecbolic or abortifacient drugs; that is, medicines which, by exciting uterine contractions, cause the expulsion of the contents of the uterus. Almost every drug in the pharmacopœia has been, at times, employed for this purpose. The drugs usually employed belong either to the class of emetics, purgatives, diuretics, or emmenagogues. It should be remembered that most of the drugs which have been administered for this purpose are not only uncertain in their action, but owing to the manner and amount administered, are usually extremely dangerous.

Violent exercise or brutal violence employed in a general manner, such as copious general bleedings, rolling down hill or downstairs, long walks, etc., have been much resorted to, and often without effecting their purpose. Over-tight lacing has also been adopted for the same purpose. It is needless to remark that the more violent of these means are almost as apt to cause death as abortion.

Abortion is frequently induced by mechanical injuries to the uterus or its contents. This is usually accomplished by rupturing or piercing the membranes by some mechanical means, such as the uterine sound, catheters, injections of water, etc. The destruction of the ovum or the rupture of the membranes of course arrests gestation; but after the membranes are pierced some days may elapse before the expulsion of the uterine contents; and as in the case of abortifacients so with mechanical means, the most violent at times fail to produce the desired effect.

**The evidence derived from the. examination of the
woman during life** is not unfrequently negative. Very
much will depend upon the period of gestation at which
the crime was committed; if committed at an advanced
period of gestation, what has been said in a previous
chapter upon delivery is here applicable. The symptoms of abortion during the earlier periods of gestation
are of an exceedingly evanescent character and may be
simulated by menstruation. The signs of abortion in
the living are commonly stated to be a relaxed condition of the vulva and passages; patulousness of the *os
uteri*, presence of the lochial secretion, and the characteristic smell common to puerperal women; distention of
the breasts; general anæmic condition; excitement of
the pulse, and dryness of the skin. The *os uteri* may be
lacerated, and there may be marks of violence on the
uterus and vagina, depending of course, upon the period
of gestation at which the abortion was committed.

If the abortion happened naturally at an early period
of utero-gestation, the signs usually found may be very
slight or altogether wanting. After the third month,
the insertion of the placenta may be detected by a rough
place on the inner uterine wall. In making a post-mortem, punctures, lacerations, and incisions of the uterus
should be carefully looked for, and care should be taken
that the uterus is not wounded in making the post-mortem, although it is not usually difficult to distinguish
wounds made before from those inflicted after death.
The stomach and intestines, bladder and kidneys should
be particularly examined for marks of irritant poisoning.
It should be remembered in this connection that the
uterus of a woman who has died during menstruation, is

thickened and presents a swollen condition of its mucous lining, and is generally hyperæmic.

As to the examination of substances expelled from the uterus, if a fœtus is found it should be carefully examined for wounds, etc., and an endeavor made to determine whether it was born alive, its age, etc., in the manner pointed out in the chapter on INFANTICIDE.

If any instruments or drugs are found in the possession of the accused, which are supposed to be the means used in producing the abortion, they should be carefully marked and preserved so as to be capable of future identification, and chemical or microscopic examination, if necessary.

CHAPTER XI.

INFANTICIDE.

THERE appears to be some difference of opinion among continental jurists as to the meaning of the term "newly born;" but according to the English and American law, no importance is attached to these distinctions, as it is equally murder to destroy the life of a child which has been completely born, according to the sense given that term in the last chapter, as it would be feloniously to kill an adult. In order, however, to constitute the crime of murder at the common law, it is necessary that the whole body of the child should have been born into the world prior to the extinction of its life. It is not necessary that the umbilical cord should have been severed nor that the child should have breathed, if it otherwise had life and independent circulation. The full period of gestation need not have passed. If a person intending to procure abortion causes the premature expulsion of the child, whereby it dies after birth from this premature exposure, this has been held to be murder. Cutting off the head, however, before the birth of the body, is not murder at the common law.

The questions which will ordinarily be propounded to the medical jurist in connection with the crime of infanticide are thus stated by Dr. Ogston : —

1. Has the prisoner been recently delivered ?
2. Was the child mature ?
3. Was it the child of the prisoner ?
4. Was it dead or alive at its birth ?
5. If alive, what was the cause of its death ?

The first inquiry has been already considered in the chapter on DELIVERY. The second inquiry will be considered under the head of LEGITIMACY AND PATERNITY: The solution of the third question will usually depend upon the application of the ordinary rules of evidence. Much light will be thrown upon the determination of this question from a comparison of the state of the child with that of its alleged mother, keeping in mind the facts already stated in the chapter on DELIVERY. It is important in questions of this kind to fix the age of the infant, or the time it may have survived birth ; and the date of the child's birth is to be settled, first, by ascertaining the date of its death, — that is, whether it died in the maternal passages, or if after birth, how long afterwards ; and secondly, by determining the period which has elapsed after its death. The indications of the death of an infant *in utero* have already been to some extent considered. Indications of its death in the maternal passages, or at or immediately after its birth, will be considered further on in this chapter.

The determination of the time the child has survived its birth rests upon the succession of changes in its body following birth, necessitated by the new physiological relations in which it is placed after leaving the uterus ; and this inquiry need not extend farther than a few days after its birth ; as, if the child has survived even a few hours it can rarely be destroyed without leaving clear testimony of its having been born alive. The changes above referred to are invariable in their occurrence, and when time has been allowed for their appearance are, except as otherwise below indicated, sure indications of the child's having survived birth. These changes are as follows : —

1. Expulsion of the meconium, which is a dark-green or olive-colored matter of a pulpy consistence, and which accumulates in the intestines of the foetus in the later months of intra-uterine life, and is discharged occasionally at birth or at periods varying from a few hours to a day or even more after birth ; it is sometimes found in the *liquor amnii*, and in breech presentations is sometimes expelled before the delivery of the head, so that the discharge of a portion of this matter cannot be regarded as conclusive evidence that the child was alive at the time of its expulsion.

2. The fall or dropping-off of the umbilical cord. In the infant after its birth, whether alive or still-born, if the death has not long preceded the birth, the cord is fresh, firm, bluish, rounded, more or less spiral, and more or less plump. The first change in the living infant after the division of the cord is the shrivelling of the part attached to the infant, commencing at its cut extremity and proceeding to the point of attachment to the abdomen. This portion of the cord then softens, while around its point of attachment there is a marked reddening or injection of the abdominal integuments. The cord next dries up, becoming at the same time brownish from its summit to its base and more or less translucent, and now presents a flattened or ribbon-shaped appearance, while through its parchment-like

walls the contracted blood-vessels within may be seen. This desiccation usually begins on the first or second day after birth, and is completed by the end of the fifth day. In the dead child, the cord, instead of a brownish, generally assumes a grayish hue. Its investing membrane forms a pulpy pellicle; it loses its previously spiral form and its vessels have not sensibly diminished in size. It is to be remarked, however, that some of the usual changes in the living cord may be produced accidentally or artificially in the dead child: thus, its parchment-like condition has been imitated by drying, or submitting it to pressure. Its spiral condition is not invariably met with in the child which has survived its birth. The separation of the cord from the body usually takes place the fifth or sixth day, sometimes a little earlier or later. During this process, the base of the cord slowly ulcerates, the umbilical artery being first divided, and somewhat later, the vein; this ulceration, with oozing of sero-purulent matter, and the existence of inflammatory appearances after the detachment of the cord (that is, up to the tenth or fourteenth day, at which cicatrization occurs), being a vital process, is of course clear evidence of live birth. In the natural fall of the cord the membranes are divided circularly and cleanly without any detached fibres, which is seldom the case when the cord has been forcibly torn away; the separation of the membranes also naturally precedes the division of the vessels. Where the cord has been violently torn off the raw and bloody edges of the part sufficiently distinguish this from the cicatrization after its natural separation. When decomposition has taken place, or when the cord has become dry, it is not always easy to determine whether its division was by a sharp or by a blunt instrument.

3. The obliteration of the internal vessels peculiar to the fœtus, namely, the umbilical arteries, umbilical vein, the *ductus venosus, ductus arteriosus,* and *foramen ovale.* The older writers on medical jurisprudence state that the obliteration of these vessels and of the *foramen ovale* occurs in a particular order and at stated times after birth, — the umbilical artery closing in from twenty-four hours to three days; the umbilical vein a little later; the ductus arteriosus in about a week; and lastly, the *foramen ovale.* Modern investigation, however, shows that their obliteration occurs, both as to time and order, in a very indeterminate manner, and accordingly little reliance can be placed upon their condition as an evidence of live birth.

4. Another sign of the child's age is the desquamation of the cuticle. This should not, however, be confounded with the peeling off of the epidermis from putrefaction after death. The desquamation of the cuticle, properly so-called, has not, says Dr. Ogston, been observed to take place sooner than twenty-four hours after birth. It usually commences at the child's abdomen and extends successively to the chest, armpits, groins, back, the extremities, the feet and hands. The shedding is usually in the shape of scales, rarely powdery dust, and may be expected to be complete in healthy infants in from thirty to sixty days; but in weakly or diseased children it may continue for an indefinite period. Where it occurs at all, Dr. Ogston considers that the child must have lived at least one day.

The series of changes above noted are regarded as only approximate in point of time.

The evidence of development derived from the state of ossification of the bones will be considered in the chapter on PERSONAL IDENTITY.

In considering the question **whether the child was dead or alive at its birth,** it is well to bear in mind the natural causes of death, and those which may result from violence not criminally employed for the purpose of causing the death of the child. The consideration of the natural causes of death is beyond the scope of this work; it may be stated, however, that among those causes are congenital debility, various diseased conditions of the vital organs, protracted labor, hemorrhage, death of the mother before the delivery of the child, fatal hemorrhages from various parts of the newly-born child, the most common of which is hemorrhage from the funis or navel, as where the cord is not properly secured by ligatures, etc. It should be remembered, however, that while fatal hemorrhages may occur from an unligatured cord, such fatal results are very rare.

The question whether the child was dead or alive at its birth may be considered under two heads: —

1. The proofs of still birth. Such signs are those which go to prove, first, that it was immature; secondly, that it was born with organs in such a state of congenital malformation — or, thirdly, of intra-uterine disease — as negatives the possibility of its surviving birth; or fourthly, that it must have died in the uterus or maternal passages before time had been afforded for the commencement of respiration or any other vital changes which take place immediately after birth.

The subject of immaturity has already, in part, been considered, and will be further considered in the chapters on PATERNITY and IDENTITY. The subject of the death of the child *in utero*, and intra-uterine maceration has already been considered. The subjects of intra-uterine disease and congenital malformation more properly belong to professed treatises on obstetrics, etc.

2. The question whether the child has been born alive is, however, usually the principal point to be determined; and the principal obstacle in the way of a satisfactory determination of this point is the senseless and inhuman rule of the common law, that the body of the child must have been entirely born before murder may be committed against it. This, however, is a question addressed to the legislature, and the medical jurist can only be expected to show that the infant died by violence or wilful neglect at the time of or subsequent to its leaving the uterus; and that but for such violence or neglect, there was nothing to prevent it from continuing to live in the external world. While it would no doubt be murder to kill a newly-born infant, born living, but which had not as yet breathed, with the exception of those cases in which the injuries are of such a character as would of themselves destroy life, and could only have been inflicted after the child had been born, it will be very difficult, if not impossible, to prove the commission of such crime. In such a case, the entire absence of those changes in the organs of respiration and circulation which distinguish extra-uterine from intra-uterine life, will render it impossible for the medical jurist to decide by mere inspection of the body whether or not the child has been born alive. In such

a case the proof of the crime must depend upon other evidence. Practically, therefore, the only cases which the medical jurist will be called upon to investigate are those in which respiration has been established prior to the death of the child. The evidence in such cases may be considered under three aspects : —

a. The evidence afforded by changes in the organs of circulation.

b. In the digestive organs.

c. In the respiratory organs.

(*a*) The signs of extra-uterine life derived from the **organs of circulation** are the obliteration of the umbilical vessels, the *ductus arteriosus*, the *ductus venosus*, and the *foramen ovale*, due to the new course taken by the blood on the establishment of respiration; but enough has been said upon this subject in a previous part of this chapter.

(*b*) The evidences of extra-uterine life derived from the **examination of the alimentary canal** are as follows : Food or medicine found in the child's stomach are of course conclusive upon the question of live birth. The food most likely to be found will be milk, farinaceous articles, or sugar. In some cases opium in some form has been found in the stomach.

In instances of drowning and smothering, the fluid or other substances in which the child has been immersed may be found in the stomach. According to Tardieu the presence of air bubbles in the glairy mucus generally found in the stomach, which he believes can only have arisen from the swallowing of saliva and mucus collected in the mouth and throat, and which is aerated by respiration, — the process, in his opinion, requiring the period of only a few minutes (from ten to fifteen at most), — is evidence of live birth.

As to the evidence afforded by the presence or absence of meconium, see *ante.*

(c) The proof of live birth principally relied upon, however, is that afforded by the series of **changes in the respiratory organs.** The time required for effecting the changes below-described varies with the condition of the infant, from a few minutes in the mature, healthy, and vigorous infant, to two hours or even more in the case of premature, weakly, or diseased infants, although cases where so long a time is required are exceptional.

Previous to the admission of air to the lungs, the lungs, on opening the chest, do not appear to fill the thoracic cavity, but occupy a comparatively small portion of the chest; the pericardium occupies a prominent place in front, and the heart is quite uncovered by the lungs, which are placed laterally and posteriorly. When, however, the lungs have been expanded by the admission of air, the pericardium is nearly covered by them, and the convexity of the diaphragm is lessened. The volume of the lungs, however, is not of itself a conclusive test of respiration.

The tissue, consistence, feeling, and color of the lungs are important to be considered in this connection. Lungs not penetrated by air have sharp, well-defined; incurved margins, are dense and fleshy, and do not crepitate when handled or cut, and no air can be expressed from their cut surfaces under water, and little or no blood is contained in their tissues. Their color varies from a chocolate hue to that of the healthy adult liver. On the other hand, lungs which have been inflated with air, whether naturally or artificially, feel spongy and light, and crepitate on being cut or handled. Although their margins appear sharp, a close examination will show that they are really rounded or projecting in tongue-like processes; they appear vesicular, and when the inspiration has been natural, blood may be squeezed out of them. By the entrance of air into the fœtal lung upon the establishment of respiration, its previous liver or chocolate color is changed to a more or less lively rose-red. This change of color, however, is said by Dr. Taylor not to be an invariable consequence of the child's having lived after birth. Lungs that have inspired air, whether naturally or not, very generally exhibit near their margins bright red stripes or patches; but in the lungs of children which have breathed naturally there are, according to Dr. Ogston, further visible on their surfaces, defined patches of bright red, relieved by the dark-purplish intervening insular spots, which form the ground tone of the whole. This last mottled or marbled aspect is considered by Casper and other continental authorities as a sure test of natural respiration; although, according to Casper, its absence does not lead to the opposite conclusion.

The so-called static test, by comparing the absolute weight of the lungs with that of the body of the infant (although fœtal lungs on the establishment of respiration necessarily become heavier), does not seem to possess much practical value.

The hydrostatic test, however, possesses great value, and in the hands of a careful inspector can rarely lead to an erroneous conclusion. It must be remembered, however, that this test is a test of respiration and not of

live birth in the legal sense of the term. In making
this test, after securing the large vessels prior to their
division, the heart, lungs, and thymus gland should be
removed from the chest, and the whole immersed in a
vessel sufficiently large to permit them freely to float or
sink, filled preferably with rain or river water at a temperature of about 60° Fahrenheit. The inspector should
note whether the parts collectively swim or sink, and if
the latter, whether the viscera reach the bottom quickly.
or slowly. The lungs should then be detached from the
heart and thymus gland, and separately tried in the
same way; they should afterwards be cut into fragments
and examined in detail in the same manner, and the
fragments should afterwards be compressed so as to expel the air as far as possible, and once more tested in
the same manner. The general conclusion derived from
this test is, that if the lungs swim, the infant has
breathed ; and that if they sink, it has not breathed.

Gas in the lungs resulting from putrefaction may be
distinguished from that resulting from respiration by.
the fact that the former collects in large bubbles, may
be easily expelled by pressure, and possesses a putrid
odor; whereas the contrary is the case with gas in the
lungs resulting from respiration, in which case no degree
of compression short of that which will entirely break up
their tissues will cause lungs inflated naturally to sink.

The objection to this test, that the lungs of the still-
born infant may be inflated artificially so as to cause
them to float independently of natural respiration, does
not seem to be in practice a valid one. Numerous attempts have been made by medical jurists to expand the
lungs of still-born infants artificially, and have almost

without exception, entirely failed. The quantity of air which may thus be introduced into the lungs has been found to be inconsiderable; and even this, as it seems, may be readily forced out by compression. Considering the difficulty of making such inflation, the ignorance of anatomy of alleged criminals likely to resort to such inflation, and their want of proper instruments and the skill to use them, the objection does not seem valid. This examination should, however, in order to be entitled to its full weight, be made by an inspector having a competent knowledge of pathological conditions which may be found in the lungs of the new-born infant; for it is possible that the condition of the lungs may be so changed by disease as to render them specifically heavier than water. Such diseases are tubercle, scirrhus, pulmonary œdema, sanguineous congestions, and perhaps other conditions which, however, will be clearly manifested if the examination is made by a competent observer.

The cases of respiration before the entire birth of the child (*vagitus uterinus vel vaginalis*) which have already been referred to do not, when properly considered, invalidate the previous conclusions; for, as we have already stated, the hydrostatic test is a test of respiration, and not of live birth in the legal sense of the term. These cases are, moreover, very rare, and none of those recorded were instances of unassisted or solitary labor; in all of these cases, either instruments or the hands had been introduced, so as to permit the entrance of air into the vagina or uterus.

Besides the above three tests, another test of live birth has been suggested by Dr. Wreden, of St. Petersburg. According to him, the gelatinous substance which fills the middle ear of the unborn infant gradually disappears and is replaced by air after the establishment of respiration, and is never encountered in the child which has lived for twenty-four hours, although it does not entirely disappear during the first twelve hours. Dr. Wendt of Leipsic, and Dr. F. Ogston, Jr., confirm the conclusion of Dr. Wreden, that this substance can only be expelled by the establishment of full respiration; but Dr. Ogston found that the time of its disappearance varies from a few hours to two or three weeks.

5. Lastly, we come to the consideration of **the causes of the infant's death.** The evidences of death *in utero* have already, in part, been considered under the head of intra-uterine maceration. Various diseases may cause the death of the fœtus before its birth, including diseases communicated directly by infection from the mother, such as scrofula, small-pox, etc., or, more remotely, from the father through her, as in syphilis, which is a very fruitful cause of the death of the fœtus. The death of the infant before birth may also be caused by mal-formation arresting development, atrophy, tuberculosis, etc. ; it may also be caused by falls, blows, pressure, etc. The consideration of the various diseases, etc., which may cause the death of the fœtus belong more properly to professed treatises on Obstetrics. See also *ante*, chapter on ABORTION.

The death of the child may also occur in a variety of ways during labor, — as from natural feebleness of constitution, imperfect development, malformation, prematurity, etc. Tedious labor is a fruitful cause of death. Death of the child during delivery, where the labor has been protracted and difficult, is most commonly from cerebral hyperæmia, *apoplexia neonatorum*. In these cases we may have congestion of the membranes of the brain and of the cerebral sinuses, and extravasations of blood between the pericranium and occipital aponeurosis, beneath the pericranium, or within the cranium; and these appearances may be so marked as to account for the death of the infant, or so slight as to be compatible with its recovery. A not uncommon consequence of tedious labor, whether the infant survives or not, is what is called *caput succedancum*, which is a sero-sanguinolent effusion under the pericranium, giving rise to a diffused swelling at the seat of the effusion, located at the parts of the head which are presented, most frequently over one of the parietal bones. Sub-pericranial effusions of blood, called *cephalhæmatomata*, occasionally result from protracted labor, and are compatible with survival after birth; they usually appear after birth, increase in the course of two or three days from the size of a half-marble to that of a chestnut or half of a hen's egg, remain stationary for a few days, and disappear slowly from a month to six weeks afterwards, leaving for a time a slight elevation of the skull at the part. Prior to the absorption of the clot, a fibrinous exudation is poured out around the detached edge of the pericranium, during the subsequent absorption of which a process of ossification sometimes sets in, converting the fibrinous ring into an osseous ridge, while that part of the cranium over which the clot has been situated is roughened by the formation of new bone on its surface.

Fractures of the skull may occur during labor in consequence of a relative disproportion of the head to the

pelvis, or of the deformity of the latter. They are, however, of very rare occurrence, and in distinguishing such
fractures from felonious injuries to the child, it should
be remembered that the amount of violence done to the
head by the prolicide is usually much greater than is
encountered in cases of fractures during labor.

It should be remembered in this connection that the
long bones of the extremities of the child are sometimes
broken or dislocated *in utero* or during delivery.

Children sometimes perish during labor from the **premature arrest of the foetal circulation,** which may arise
from the detachment of the placenta, prolapse of the
cord in head presentations, or compression of it by the
head, or by the body or the head in foot presentations, and from the cord's becoming twisted around the
child's neck or limbs, and thus suffering compression. Under such circumstances the arrest of the circulation leads to instinctive efforts on the part of the
child to breathe, and therefore to death by a species of
asphyxia.

In cases where the cord has become twisted around
the child's neck during labor, thereby causing death by
asphyxia or coma, in the great majority of instances no
mark is made on the neck by the cord; exceptional
cases are, however, recorded in which furrows and discolorations of the skin have been found at the points
which have suffered compression.

Among the natural **causes of the death of the infant
subsequent to delivery** may be mentioned certain malformations incompatible with extra-uterine life; and
immaturity hindering its surviving birth.

The infant may die after delivery from natural feebleness, or from the presence of disease which commenced before birth, such as tuberculosis, syphilis, etc.

There are recorded instances, also, in which the child has perished after delivery from occlusion of the mouth and nostrils by the membranes, where the child and membranes are suddenly discharged *en masse.*

Experiments by Dr. Casper also show conclusively that death may be occasioned in consequence of a fracture of the skull by the child's falling from the genitals of the woman in a standing posture, in case of a sudden or unexpected delivery.

While it may be that the infant may perish after delivery from the rupturing of the umbilical cord and consequent fatal hemorrhage in cases of a sudden expulsion and consequent fall of the infant, there do not appear to be any authenticated instances of such fatal hemorrhages.

A child may be smothered in the bedclothes or by being overlaid, either accidentally or criminally; but there are no means of distinguishing the one case from the other. The child may also after birth perish by its face falling into the mother's discharges, or by being accidentally dropped into a privy when labor has come on suddenly at stool.

Criminal practices against the life of an infant are usually perpetrated at the conclusion of labor. It will often be impossible in practice to distinguish those cases where the death of the child has arisen from want of suitable warmth or nourishment, absence or unsuitability of the ligatures of the cord, etc., where it is accompanied with a criminal intent, from cases where there is no criminal intent.

The life of the infant has been destroyed by punctures of the fontanelles, of the orbit, nucha, twisting of the neck after delivery of the head, strangulation, and wounds inflicted through the several inlets and outlets of the body. The body of the infant whose death is suspected to have been caused by criminal violence should be carefully inspected for such injuries. Death may also be caused by strangulation, blows or injuries about the head, incised wounds, drowning, suffocation, smothering, etc. The appearances characteristic of death thus caused will be found discussed in the chapters upon WOUNDS, DROWNING, STRANGULATION, etc.

In a case of death by drowning, unless the child had lived long enough to have established respiration, this mode of death would leave no traces on the body.

In cases of suspected poisoning of new-born infants, it should be remembered that the internal surface of the œsophagus is often injected in new-born children after death, such injection assuming the form of ramification of vessels or of longitudinal striæ, which might be mistaken for the effects of certain irritant poisons.

Ulcerations of the stomach, attended with the collection of a brown or blackish bloody fluid, are sometimes also discovered, which might give rise to a suspicion of poisoning. Similar appearances have also been found in the intestines, thus affording sufficient data by which to account for the death of the child from natural causes.

We cannot better close this chapter than by giving Dr. Ogston's method of making an inspection of the body of an infant suspected to have been the victim of foul play. The medical jurist, says he, ought at the outset to make himself acquainted with all the particulars of the case, — the place where the child was found, the circumstances attending its removal thence, and the facts relative to the suspected party if suspicion rests upon any one. He should, if possible, visit the place where the body was discovered; he should then examine the wrappings of the child, if there are any; the marks upon them, if any; the thread with which they had been sewed; character of the ligature, if any, around the cord, etc. In this way the mother may sometimes be discovered.

These preliminaries having been attended to, the examiner should proceed to the **external inspection** of the infant's body, noting —

1. Its general conformation, and especially any defect or vice which might effect its viability.

2. Its degrees of freshness or putridity.

3. The color of the skin.

4. The degree of adhesion of the cuticle and of the nails.

5. The extent of the saponification if it has commenced, or of the emphysema if gases are generated under the skin, etc.

6. The natural openings should be examined as to their being pervious or otherwise, with the discharges which may have proceeded from them.

7. If any punctures, incised wounds, contusions or ecchymoses exist, their situation, extent, and depth should be especially noted.

8. The body should be weighed and measured, observing with what point of the abdomen the centre of its length corresponds.

9. The state of the navel should be particularly attended to, observing if any part of the cord remains attached to the belly, whether it is fresh or shrivelled and dry, or free or not from inflammation and vascularity; whether it is spare or plump; whether it has been tied or not; if tied, at what distance from the navel; the degree of torsion of the remains of the cord; its translucence, the volume and course of its vessels, if they contain blood or if any can be expressed from them; whether it has been cut through or torn, and if the former, whether with a blunt or sharp instrument.

10. If the placenta is found, the divided ends of the cord should be compared to see if they correspond, and the length of the remains of the cord, both placental and fœtal, should be measured.

11. The sebaceous coating, if present, should be looked for chiefly in the arm-pits, groins, and hands; its presence shows that no care had been taken by washing for the child's preservation.

12. The hairs should be noted as to their color, length, and quantity.

This external general inspection completed, the examiner should proceed to the **special inspection**, beginning with —

1. The head, — ascertaining its form and dimensions in different directions; carefully removing the hairs and examining the scalp; dividing the scalp by a crucial incision, or from ear to ear; ascertaining the state of the internal surface of the scalp, of the bones and sutures, and the size and appearance of the fontanelles; removing the skull-cap in the usual direction with a pair of stout scissors; examining the brain *in situ*, and

removing it along with the *medulla oblongata* in order
to inspect the top of the spine and the base of the
skull.

2. The front of the neck should then be attentively
looked at for any grooves caused by ligatures, ecchymoses or abrasions, after which the larynx and vertebral column may be inspected.

3. The cavity of the mouth should now be laid open
to ascertain if any plug had been introduced and left
there, or if any traces of irritants or corrosives appear.

4. To ascertain the state of the chest, the clavicles
should be divided with a pair of scissors, avoiding the
subclavian vessels, and afterwards the cartilages of the
ribs, folding down the sternum and leaving it attached
to the abdominal parietes. The situation and volume
of the pectoral viscera can now be noted, as well as the
state of fulness or vacuity of their vessels. It should
now be observed whether the lungs cover the pericardium or lie unexpanded and deep in the chest;
whether their margins are sharp or rounded; what
is their color and consistence; whether uniform or
mottled; whether their capillaries are injected or not;
whether they are fresh, emphysematous, or putrid.
Their appearance can be compared with that of the
thymus gland and the appearance of that gland noticed.
The pericardium should now be opened and its appearances, externally and internally, examined; the thymus
gland turned up, the left lung being pushed to the right,
and the state of the *ductus arteriosus* ascertained. After
the application of double ligatures to the *venæ cavæ*, the
aorta, pulmonary artery, and the trachea, the lungs and
thymus gland should be removed from below upwards,

and the hydrostatic test applied. The heart should next be opened to ascertain the quantity and distribution of the blood in its interior and the state of the *foramen ovale*. The heart, lungs, and thymus gland should then be separated from each other, — after securing their connecting vessels with double ligatures, — the lungs weighed, and their weight compared with that of the body. Note whether they sink or float — whether in whole or in part, and whether buoyantly or not — when thrown into water; and, after cutting them into fragments, what portions sink and what float; if the former, whether they are sound or diseased, and what the disease, if any. On handling and cutting the lungs, it should be observed whether they have a spongy or solid feeling, and whether they crepitate or not. By compressing the fragments in air, the amount of blood or other fluids which they may contain can be ascertained; while by doing so under water, the presence or absence of air can be discovered; and if they contain air, the size of the air bubbles given out will assist in the discrimination as to whether it was contained in the air cells or in the subpleural or interlobular areolar tissue, and their odor if fetid will serve to characterize putrid lungs.

5. The best mode of opening the abdomen so as not to interfere with the umbilical vessels is to remove the sternum at the base of the chest, to carry an incision from the xyphoid or ensiform cartilage to a point a little above the umbilicus, to prolong the incision downwards and outwards at a little distance from the navel to the anterior superior spinous process of the ilium, and thence across the pubes to meet the continued incision on the

opposite side. The examiner should now search the abdomen for effused blood and sanguinolent serum or other effused fluids; for marks of putrefaction; for ruptures of the liver, or ecchymoses indicative of effused blood in its interior; for rupture or softening of the spleen. The state of the umbilical vessels should be noted, whether open or contracted, full or empty; the state of the stomach; whether it contains milk or other alimentary matters, or mucus containing air bubbles; and in what part of the intestines the meconium, if present, is to be found, or if they contain air. The exploration of the bladder, kidneys, and genitals, external and internal, complete this part of the examination.

6. The spine should be explored throughout.

7. To complete the inspection, it only remains for the examinator to ascertain whether any ecchymoses exist on the trunk or limbs, for this purpose making free incisions in the latter; and to turn his attention to the bones in order to ascertain the state of their development.

See the chapters on MEDICO-LEGAL INSPECTIONS, LEGITIMACY AND PATERNITY, and PERSONAL IDENTITY.

CHAPTER XII.

DEFLORATION AND RAPE; SODOMY.

Definitions, etc. — By "defloration" is meant the act of depriving a female of her virginity; and by "rape" is meant carnal knowledge of a woman by force and against her will. Cases of the former will seldom come before the medical jurist.

The age of consent by the common law is fixed at twelve years; this age has, however, been variously changed by statute to ages from ten to fourteen years. Carnal knowledge of a child under the age of consent also constitutes the crime of rape, a child of such tender years being considered incapable of giving a legal consent. According to the doctrine of the common law, a boy under fourteen is incapable of committing rape whatever may be his physical capacity. In the State of Ohio, however, this doctrine has not been adopted; in that State a boy under the age of fourteen is *prima facie* incapable of committing this offence, but on proof of the existence of such capacity, he may be convicted of this crime. The same point has also been ruled in the Supreme Court of New York, but does not seem to have been adopted elsewhere.

By the law as it at present exists both in England and in this country, emission is not necessary to the complete commission of this offence. Penetration is, however, necessary; but both the English and the American courts hold that nothing more is required than *res in re*, without regard to the extent of penetration; and according to the prevalent doctrine, mere vulvar penetration is sufficient,— the fact of the hymen's not being ruptured being only presumptive evidence against penetration.

Without entering into the consideration of the legal aspects of this offence, which more properly belongs to a treatise upon the criminal law, it is proper in this connection to consider the subject under three heads :

1. Violation of the female under the age of puberty.

2. Violation of the female after puberty but prior to her having otherwise had sexual intercourse.

3. Violation after puberty and where the female has been accustomed to such intercourse.

Before entering into the consideration of these three topics, it may perhaps be well to state **the physical signs of virginity.** No female should be examined for the purpose of ascertaining whether or not she is a virgin during the period of menstruation. The physical signs indicating virginity are, —

1. The presence of an intact hymen, which is a membrane stretched across the entrance to the vagina, in which there is an opening towards the orifice of the urethra. This opening may assume a variety of forms, as that of an irregular, circular diaphragm broken at its upper third or in some cases perforated by a central opening; or, which is its most common appearance, it may consist of a semi-circular fold of integument stretched across the lower border of the vaginal orifice.

the free border concave and notched and its extremities losing themselves in the *labia minora*. It sometimes assumes other forms, such as a complete septum pierced by numerous minute openings; and there are recorded instances of its being imperforate and completely blocking the entrance to the vagina. Its normal condition is such as to permit the passage of the finger into the vagina without injuring its border of mucous membrane; and there would seem to be no doubt that if this aperture was slightly larger than usual, and the male organ small, repeated intercourse may be had without rupturing it or without other change than an increased dilatability of the aperture, or perhaps minute tears or indentations on its free border. In some females the hymen may be in so relaxed a state, by the discharges of leucorrhœa or otherwise, as to diminish its liability to rupture by sexual intercourse. If, however, the aperture is small, it will almost always be ruptured by the first coitus. There are recorded cases, however, where an unruptured hymen has continued throughout pregnancy and has not been ruptured until delivery. Dr. Lusk records the case of a young woman, nineteen years of age, who possessed a perfect hymen, the opening of which was of the ordinary size, yet so distensible was its tissue that a one inch speculum was repeatedly introduced for purposes of examination without in the slightest degree affecting its integrity. There is said to be in Meckel's museum at Halle, a specimen of the female genitals where the hymen is perfect although the woman had given birth to a seven months' child. Dr. Lusk states that in his experience, however, in the examination of young, nulliparous prostitutes for uterine disorders, he has always found a torn hymen, but in no case *carunculæ myrtiformes.*

Although the existence of the hymen in any case has even been doubted by some writers, there can, we think, be no doubt of its constant existence at some period. With respect to this subject Casper makes the statement that when a forensic physician finds a hymen still preserved, even its edges not being torn, and along with it a virgin condition of the breasts and external genitals, he is then justified in giving a positive opinion as to the existence of virginity.

While not prepared to indorse so strong an opinion as this to its fullest extent, still such a condition — especially if the aperture is small, and the membrane undilatable and of ordinary shape and structure and normally placed — is without doubt strong evidence of virginity, though perhaps not absolutely conclusive. Notwithstanding this, however, the presence of an intact hymen must not be considered as conclusive proof that rape has not been committed, especially considering the fact that the crime may be completed by simple vulvar penetration.   It should also be remembered that in children this membrane is more deeply placed than in the adult, and can only be readily seen in them on the forcible separation of the thighs.

Although the presence of an unruptured hymen in a female arrived at puberty is very strong evidence of virginity, it may be destroyed by surgical operations, by a medical examination of the genital organs, by self-abuse, and by various other forms of mechanical violence. It has even been alleged that riding, dancing, and leaping may in exceptional cases destroy it, —

allegations, however, which, considering the location of the hymen, are in our opinion to be received with considerable scepticism.

2. The absence of the *carunculæ myrtiformes* is usually regarded as evidence of virginity. Two or more of these are vaginal; the others — which are hymeneal, and are the remains. of the ruptured hymen — only prove the rupture of the hymen, and furnish no evidence as to the cause of its rupture.

3. The entirety of the *fourchette, fossa navicularis*, and posterior commissure of the labia, are regarded by. some as another evidence of virginity. Of these it may be said that while they seldom survive the first labor at term, they are not, as a rule, affected by sexual intercourse unless under circumstances of considerable violence; so that while destruction may be evidence of prior delivery or at least of the passage of some large body through the genital passages, their presence furnishes very little evidence of virginity.

4. A narrow and rugose state of the vagina is sometimes considered a mark of virginity. It not unfrequently, however, exists in young, healthy married women prior to child-bearing, and is not always obliterated after a single confinement at an early age. Mere sexual intercourse, unless under circumstances of great violence, would not alter its condition; while on the other hand leucorrhœa or other pathological conditions might destroy this condition in females who had never had sexual intercourse.

5. A plump and elastic condition of the breasts, with slight development only of the nipples, is mentioned by medical jurists as another evidence of virginity. The

areola is altered by conception, but not by mere sexual
intercourse; and although the breasts may be slightly
affected by constant intercourse, and considerably so by
advancing age and feeble general health, the evidence
of virginity afforded by their general condition would
seem to be very slight.

6. The integrity of the *perinæum*, while always found
in virgins, is not affected by sexual intercourse unless
accompanied by great violence; and is therefore of very
little value as an evidence of virginity. It is, however,
almost always slightly lacerated in first labor.

7. In healthy virgin females before the decline of life,
the labia are usually bulky, smooth, vermilion-hued,
elastic, and in contact with each other, concealing the
orifice of the vulva entirely; and the nymphæ are
smaller in proportion in a virgin state of the genitals,
and lie more concealed by the labia when the thighs
are nearly in contact. Nevertheless, indulgence in
libidinous desires and manipulations would doubtless
produce in them a similar effect to that produced by
sexual intercourse. The normal size and condition of
the clitoris affords a presumption of virginity; and its
enlargement, facility of erection, and the laxity and
mobility of its prepuce, raises a presumption of non-
virginity. But here, again, it should also be remembered
that the practices above referred to may produce the
same as or greater changes than sexual intercourse.

The concurrence of all these physical signs of virginity affords strong evidence of this condition; and yet
it behooves the medical jurist to be very careful in ex-
pressing a decided opinion upon this subject; for there
is a recorded instance of a common prostitute's preserv-

ing so many of the signs of virginity that a very able
physician after a careful examination was unable to
determine whether or not she was a virgin.

1. Coming to the first of the divisions above indicated, namely, **violation of the female under the age of
puberty**, in such cases extensive marks of local injury
should be looked for, and in their absence it cannot be
admitted that complete coition has taken place. This
will be evident when we consider the undeveloped state
of the sexual organs in the female at this age, and the
disproportion between the adult penis and the narrow
canal of the vagina. Instances are sometimes met with
of violence produced in this way on children so serious
as to prove fatal to life. In one case recorded by Dr.
Ogston, a post-mortem examination disclosed the external organs of generation and the perinæum torn and
violently inflamed, the vagina torn away from the
uterus, and a large rent in the peritoneum, with bloody
fluid effused into the abdominal cavity. Injuries of all
degrees may be inflicted between this extreme and the
other extreme of mere vulvar penetration without entrance into the vagina and injury to the hymen; in the
latter case there may be no trace of violence, but the
crime may be complete, in the legal sense, in the entire
absence of marks of injury about the external genitals.
In the simpler cases there may be slight irritation of
the vulva, characterized by heat and moderate redness
of the parts; more frequently, however, there will be
swelling and contusion of the labia, intense redness of
the hymen and vaginal outlets, and the whole of these
parts will be so painful as to make any examination of
them difficult, if not impossible. Excoriations, super-

ficial erosions, and even real ulcerations are not unusual. In this connection it should be remembered that diseases of the genitals occasionally occur in female children which unless care is used may be confounded either with the effects of local violence or with venereal disease.   One form of such local disease in young females consists of a purulent discharge from the genitals closely resembling gonorrhœa.   A muco-purulent vaginal discharge will probably be found within a few hours after a rape upon a young girl; this discharge is not gonorrhœal but results from the inflammation arising from the irritation of connection.   It is usually at first bloody, but rapidly changes to a greenish tint, finally becoming glutinous; it is commonly attended with great smarting and with constant desire to scratch the genitals; but the existence of such a discharge, although it is important evidence, is not proof of rape.   It is a well-known fact that the majority of females, whether virgins or not, suffer at times from leucorrhœa; while in the case of female infants and young children inflammation of the vulva and of the vagina, giving rise to infantile leucorrhœa, is common.   It is generally considered that there are no certain diagnostic signs by which, under all circumstances, such discharges may be distinguished from leucorrhœa, unless it be by a pure culture and microscopic examination for the discovery of the so-called gonococcus; but in the present state of our knowledge, the specific nature of this so-called *gonococcus*, while probable, can hardly be said to be proved.

Another severe form of genital inflammation sometimes found in young children, and which terminates in a destructive and gangrenous form of ulceration is the disease called *noma*, which might be mistaken for the results of attempted violation. A further description of this disease is beyond the scope of this treatise.

Light may be thrown upon the case by a careful inquiry as to the time when the discharge first appeared; for if gonorrhœa or syphilis be clearly marked in the female at the time of the examination, and such examination has been conducted immediately after the alleged rape, it may be considered as certain that the venereal disease did not result from the alleged rape; for both gonorrhœa and syphilis have a distinct period of incubation, which in the former disease varies from some hours to three or four days or even more, and in the latter from ten to forty-four days or even longer. The existence, therefore, of a profuse discharge a few hours after an alleged rape, or of secondary or tertiary symptoms soon after, would be almost conclusive evidence that the disease was contracted prior to the alleged rape.

It must also be remembered that this disease may have resulted from other causes than intercourse, and hence if sufficient time has elapsed for the development of the disease, the existence of the disease is not very strong evidence of the commission of the crime. If an examination of the male shows that he has not either of said diseases, of course this is strong exculpatory evidence; if on the other hand he is found suffering from the same disease as the female, this may be evidence against him.

2. Coming now to the subject of the defloration or **violation of a virgin after puberty**, it may be stated at the outset that the proof of such violation by medical evidence is much more difficult than in the case of

children.   In order to establish the offence in such cases there must be evidence —

a. That previous connection with another person could not have taken place.

b. That the alleged intercourse took place at the time alleged.

c. That it was by force and against the will of the woman.

The evidence bearing upon the first point has already been considered at the beginning of this chapter, under the heading, **physical signs of virginity.**

As to the two other points, where both parties are above the age of puberty, and the penis of the male is disproportionately large as respects the female, a recent connection, when it has been fully effected, will produce effects in the female the same in kind, though less in degree, as in the case already considered.   This dispro-portion may exist where the male is in the vigor of manhood, and the female just past the age of puberty, and her sexual organs not yet fully developed.   It may also exist where both parties are of mature age, but the male is large and vigorous, and the female small or not so fully developed.   In such cases the following ap-pearances may be encountered, namely: bruising and excoriation of the clitoris and labia; laceration of the mucous membrane of the external parts, with ecchymo-ses under the membrane; vascular injection in the vicinity of the excoriations; rupture of the hymen and sometimes of the fourchette, and even excoriation of the lining membrane of the vagina within its external ori-fice.   Besides these, the vulva may be swollen, tender and painful to the touch, while the chemise may be stained in front with seminal fluid, and behind and in front with blood. These appearances may sometimes present themselves to some extent where the venereal congress has been entirely voluntary on the part of both, especially where the parties have been actuated by strong and ungovernable desire. These appearances of the genitals must be looked for early; in almost every instance they will have become obliterated by the third or fourth day, by which time the lacerations will have healed and the torn hymen be in such a state as to make it difficult to say whether it has been ruptured recently or at an earlier period.

In the cases above described the male has been supposed to be more vigorous and fully developed than the female; when, however, the parties are more nearly equal in this respect the effects of a first connection will be likely to be much less noticeable, though some of them will probably be found. When the genital organs of the female have become relaxed by long continued leucorrhœal discharges, profuse menstruation, advanced age or other causes, the signs of recent defloration will be still more faint.

It should be remembered that local injuries such as those which may be produced by first coitus, and those which may be caused by the forcible introduction of foreign bodies, cannot with certainty be distinguished; and that there are recorded instances in which such injuries have been produced by the assailant not in the usual way, but by other means than by the introduction of his penis. These questions must be solved by the ordinary rules of evidence.

The subject of **seminal and blood stains** which have

already been referred to, as well as marks of bruises, scratches, etc., will be further considered *post*.

3. Coming now to the third division, namely, **where the female is known to have been accustomed to sexual intercourse,** it may be stated that the proof in these cases must usually be made out by the application of the ordinary rules of evidence, and usually little assistance can be rendered by the medical jurist. Of course the signs of virginity and of defloration are not elements of the inquiry.

It should be remembered, however, that in these cases, where the disproportion between the bulk of the penis and the size of the vagina is very great, the appearances above-described may to a greater or less extent be produced even where the woman is not a virgin; especially may this happen where the intercourse is effected with brutality, or where several men in succession have forcibly had connection with the same woman. In this class of cases, as well as in children and virgins, bruises, scratches, or other injuries of the person will often be met with. Seminal stains will probably be found, and venereal disease may follow the intercourse. The evidence, however, to be derived from the existence of venereal disease is much less trustworthy than in the cases previously described. It is not necessary in this connection to add anything to what has been said upon this subject. In either of the last two cases pregnancy may result from the alleged rape; but in the case of married women or women living in adultery the evidence afforded by such impregnation will of course be of no value.

Blood-stains on women who are accustomed to sexual

intercourse will not be found unless the intercourse is effected under circumstances of exceptional brutality.

There are recorded instances of sexual intercourse having been had with a woman during a deep sleep, without her having been conscious of the fact. Where the woman has been accustomed to sexual intercourse, or where the sleep is pathological or induced by drugs, this is not impossible; but that intercourse may have been had during natural sleep with a woman not accustomed to sexual intercourse, or, as a rule, even with a woman so accustomed, is not to be believed except upon the most conclusive evidence. There is one recorded case where rape was committed upon a woman in a state of mesmeric coma.

The allegation is sometimes made in cases of alleged rape **that the accused exhibited chloroform** upon a handkerchief to the woman, who thereupon instantly became unconscious. Such allegations are entirely opposed to universal experience and are not to be believed. Experiments have been made to determine whether a person may be brought under the influence of chloroform during natural sleep without awakening him; in the vast majority of cases it has been found impossible of accomplishment. There are, however, a few cases in which it has been done as an experiment by medical men experienced in its administration; that it could be accomplished by one not so experienced is hardly credible.

Not unfrequently charges of rape while under the influence of chloroform have been made against dentists; and in some instances conviction has followed such charges. While a dentist who would administer such an anæsthetic without the presence and assistance of a competent medical man is, perhaps, deserving of some punishment, such charges are to be looked upon with considerable suspicion; for it must be remembered that not unfrequently women while under the influence of chloroform experience erotic dreams which have such a semblance of reality that they are with difficulty persuaded of their non-reality when consciousness has been restored.

It should be a universal law of practice with all medical men never to administer any anæsthetic except in the presence and with the assistance of another competent medical man if possible; or at least, never without the presence of some other reliable witness.

An examination of the accused may often furnish important evidence in a case of alleged rape. The subject of venereal disease has already been sufficiently considered. Scratches, especially on the face, hands, and penis, tears of the clothes, and spots of blood, semen, or dirt, should all be noted. Sometimes a laceration of the *frœnum prœputii* will be found, especially if the accused is not accustomed to sexual intercourse.

Coming now to the **stains** on the woman's linen, — they are usually of two sorts, one pale and the other colored. The colored stains are usually on the back part of the chemise, and consist principally of blood. There are two sorts: one of a deep red, rich in coloring matter equally diffused over the stain; and the second sort of a much lighter red, or rather, reddish-yellow color at the centre of the stain, and their margins having a defined reddish line. The former of these stains arises from pure blood; the latter from a sero-sanguinolent discharge, becoming gradually less and less red. The subject of blood stains and their microscopical examination will be more fully considered in another place.

Besides the reddish stains arising from blood above-described, there may be found broad diffused yellowish ones caused by urine; those of a yellow or greenish-yellow color due to feculent matter; light-yellow, diffused and stiffened patches from muco-purulent or mucous discharges, etc. Seminal stains found on the linen after coition are usually situated in front, at a point corresponding with the vulva, although they are occasionally found behind.

In the examination of a woman alleged to have been violated, the pubic hairs should be carefully examined for spermatozoa, which adhere very closely to them. In general appearance a seminal stain is colorless and stiff; by transmitted light it is of a somewhat grayish-brown tinge; when warmed it becomes of a pale-yellow tint, which is quite characteristic, and happens with hardly any other discharge, healthy or morbid. When warmed or moistened with warm water, the characteristic odor of seminal fluid is evolved. If a small portion of the stained fabric be digested in a watch-glass with a few drops of water for about ten minutes, and the water carefully squeezed out from the fabric, and to the solution a drop of nitric acid be added, the glass being placed in a good light on a piece of white paper, if the stain is seminal there will be no precipitate, but the liquid will turn to a yellow color. In pure seminal stains the guaiacum test gives no blue reaction. Where the garment upon which the stains occur is dirty or colored, the above tests are of no value; and, as a general rule, the detection of the characteristic spermatozoa by microscopic examination is the only test which in practice ought to be relied upon; and nothing should be admitted to be a seminal stain unless complete spermatozoa are found.

Human spermatozoa have a flattened, almost oval head, with a long, slender, filamentous tail; the entire length varying from $\frac{1}{500}$ to $\frac{1}{800}$ of an inch, — although some may be found that do not exceed $\frac{1}{1000}$ of an inch in length. The tail is usually five or six times as long as the head, which is about $\frac{1}{8000}$ of an inch in diameter. The shape of spermatozoa varies in different animals. Human spermatozoa are quite tenacious of life, and there are recorded instances in which active spermatozoa have been found in the vaginal mucus several days after intercourse. There is one case on record in which they were found alive fourteen days after the rape. They have a remarkable power of resisting putrefaction, and even after they are dead and the stain has become dry, they may by the aid of a microscope be easily distinguished by their characteristic shape even months after their emission.

Microscopical literature abounds with descriptions of the technique of their examination. Without entering into this subject in detail, it may be stated that the examination should be made with the power of 300 or 400 diameters, and that in preparing material for examination the stained portion of the fabric should be cut out with as little handling as possible, placed in a clean watch-glass and moistened with three or four drops of cold distilled water. After having allowed it to soak

for about ten minutes, moving the fabric about gently
with a glass rod, the water may be squeezed out with the
end of the rod and specimens of the liquid and deposits
placed on a glass slide, covered with a cover-glass and
examined.   If no spermatozoa are at first found, the
fabric may be carefully unravelled and different portions subjected separately to examination.   An aqueous
solution of methyl blue or fuchsin will often serve to
differentiate the spermatozoa.

It should be remembered, however, that spermatozoa
do not always occur in the seminal fluid, especially in
the prostatic fluid which has been ejaculated after one
or more prior emissions; and also that in some cases of
disease no spermatozoa are found.   Their absence also
is not unusual in the seminal fluid of the old and the
feeble, so that the failure to find spermatozoa in the
alleged seminal stain will not invalidate a charge of
rape ; and their discovery, upon the other hand, is at the
most, only evidence of the fact of previous sexual intercourse without determining whether it was voluntary or
involuntary.   The expert microscopist will not be likely
to confound human spermatozoa with anything else,·
There are, however, certain fungi which are said to resemble spermatozoa; their tails, however, are clumsier,
and they refract light differently.

An animalcule called *trichomonas vaginalis* is sometimes found in the vaginal mucus of females not remarkable for their cleanliness, which has been supposed to
bear some resemblance to seminal animalculæ.   The
heads of the trichomonads, however, are very much
larger than those of spermatozoa, and are, moreover, furnished with several ciliæ, while the spermatozoa have

none. Internally, moreover, the trichomonads are granular, which is not the case with spermatozoa.

The third element of the crime of rape, namely, that the intercourse was by force and against the will of the female, has already been somewhat considered in what has preceded. As a rule, independent of the marks of violence on the person of the woman, evidences of a struggle, etc., this element of the crime must be established by the ordinary rules of evidence with which the medical jurist has nothing to do. This topic may, with propriety, be left to the officers charged with the administration of the law.

Sodomy. — Sodomy, sometimes called "buggery," sometimes "the offence against nature," and sometimes "the horrible crime not fit to be named among Christians," consists of a carnal copulation by two human beings with each other against nature, or by a human being with a beast. Unlike rape, sodomy may be committed between two persons both of whom consent; it may be committed between husband and wife, between two men, or a boy and a man. By "sodomy" is ordinarily understood the offence between man and man, or between man and woman; where the victim is a boy, sodomy is usually termed "*pæderastia*."

. By "bestiality" is meant intercourse by the human kind, male or female, with an animal, male or female, other than the human kind. The term "buggery" includes unnatural intercourse both with mankind and with animals.

According to Lord Coke, emission is a necessary element of the crime of sodomy, but at present time by statute both in England and in this country, penetration

is sufficient, as in rape, without proof of emission. A
penetration of the mouth is not sodomy, neither is an
unnatural connection with a fowl, — a fowl not coming
under the term " beast."

As to the appearance of, and the effects produced
upon the criminals, it is said that if the crime has been
habitual and frequent, there will be the usual evidence
of sexual excess, and premature decay of strength, the
apparent age of the person far exceeding the real, etc.
The penis is said to be commonly found elongated; the
glans more than usually bulbous and conical, and the
urethra twisted. The natural fold about the anus of
the passive criminal, and those that radiate towards the
anus, become rapidly obliterated by repetitions of the
crime, giving the skin of the part a smooth appearance;
moreover, a peculiar funnel-like or trumpet-shaped depression or hollow of the nates is usually observed.
This appearance, however, it should be observed, may
be produced by other than criminal practices, such as
the necessity in some people of pushing back piles or
slight protrusions of the rectum forced out during defe-
cation. Where the crime is committed on one unac-
customed thereto, the passive agent, if examined soon
after, will exhibit a certain amount of bruising and inflammation, with a slight laceration of the sphincter.
There may be congestion and abrasion of the anal mucous membrane without injury to the sphincter; in
some cases seminal stains will constitute important
evidence.

Where that form of sodomy known as bestiality has
been committed, the principal medical evidence will
consist in the finding of spermatozoa on the person or

clothes, or on the hairs of the animal; or in identifying the hairs of such animal on the accused, both of which may be done by the aid of the microscope. If the hairs of the animal be found adhering to stains of blood, mucus, or semen, on the underclothing of the accused, this fact will have considerable weight.

11

CHAPTER XIII.

IMPOTENCE AND STERILITY.

QUESTIONS respecting impotence and sterility more frequently arise upon bills for a divorce; although impotence may sometimes be alleged as a defence in rape, or may be given in evidence upon the issue of bastardy or illegitimacy of a child, and perhaps in other cases.

Marriage between two persons of the same sex is void, because none of the ends of matrimony can be accomplished thereby. The subject of doubtful sex will, however, be considered in another chapter.

In order that a marriage may be valid, the parties, both of opposite sexes, should have their sexual organization and capabilities essentially complete; impotence, therefore, is everywhere considered sufficient ground for decreeing the nullity of a marriage. In countries governed by the common law, a marriage will be good if there is an adequate power of mere copulation, though in the ordinary case pregnancy cannot be made to follow; but the copula must be adequate as such or the marriage will be invalid; yet if pregnancy is actually produced by an otherwise inadequate copula, this, according to Mr. Bishop, will render the marriage unimpeachable. The doctrine settled by the weight of authority is believed to be that the ground of interference by the court in cases of impotency is the practical impossibility of connection. It is not necessary, as it seems, that there be a structural defect rendering copulation impossible; if it is impracticable that will be regarded as sufficient. It has been held in several cases that, where there is a practical impossibility of copulation, which, however, the evidence disclosed might be removed by surgical treatment which would not endanger the life of the respondent, if the respondent refuses to submit to such treatment, a decree dissolving the marriage will be granted. Where, however, there is capacity of body, but the respondent merely refuses to submit to the embraces of the complainant, this constitutes no ground for a decree. The defect, then, must either be incurable or, according to the better opinion, it must be such as practically prevents sexual intercourse; and if curable, the respondent must have positively refused to submit to treatment for that purpose. Where, however, the medical evidence shows that the malformation might possibly at great risk of life and by an operation of doubtful success, be removed, the petitioner is not bound to call upon the respondent to submit to such a risk, and this state of facts is deemed equivalent to a permanent and incurable malformation.

It appears to be settled that mere incapacity of conception where there is capacity for copulation is not sufficient ground whereon to found a decree of nullity.

The causes of impotence in the male are chiefly the following : —

1. **Extreme youth,** before the arrival of puberty, is of course accompanied by impotence. The ages of fourteen

in the male and twelve in the female are recognized by
the common law as the earliest ages at which a valid
marriage can be contracted. These ages have been
changed somewhat by statute in different States. Puberty is sometimes anticipated, there being recorded instances of puberty in boys as early as four or five years,
and in girls of the presence of menstruation almost from
birth. On the other hand, puberty is often deferred
until quite an advanced age. Mere old age cannot be
regarded as involving sexual incapacity, although the
sexual powers are usually very much diminished by
old age.

This general subject, however, will more properly be
considered in the chapter on LEGITIMACY AND PATER-
NITY.

2. The impotence which more frequently is a ground
for a decree of nullity is that which arises from congenital or other defects. The subject of hermaphrodism
will be considered in another connection; suffice it to
say here that if it exists to such an extent as to defeat
the ends of marriage, it is ground for a decree. The entire absence of the penis, whether congenital or the result
of disease or accident, is clearly ground for a decree.
When there is any intromittent organ, however small,
the medical jurist should be very cautious in pronouncing a man impotent. It should be remembered that
impregnation may occur with an unbroken hymen and
where the male organ is so short as only to be capable
of depositing semen within the vulva. It may, perhaps, be stated that a charge of impotence on account
of the abnormal condition of the male organ will not be
regarded as sustained short of evidence of the absolute

and complete loss of the penis, or of its most extraordinary development. Each case, however, as it seems, should be decided on its own facts in accordance with the general principles above stated.

An extreme degree of hypospadias or epispadias may in some instances amount to incapacity. Some other malformations, or complaints, such as the abnormal direction of the penis, its attachment to the abdomen, phymosis, paraphymosis, etc., are sometimes alleged as causes of impotency. As in most cases they may be corrected by a slight surgical operation, there would seem not to be ground for divorce in such cases.

No reported case has come to our notice deciding the question whether the absence of both testicles, the power of copulation remaining, however, is a ground for a divorce. The absence of one testicle, or the nondescent of the testicles, is clearly consistent with the ability to procreate, and there would seem to be no doubt that in such case there is no ground for a divorce. This subject will, however, be considered further in the chapter on LEGITIMACY AND PATERNITY.

3. Impotency may be caused by **disease of the male sexual organs**, such as syphilis, cancer, or tuberculosis. Masturbation may be a cause both of sterility and impotency.

Injuries to the head and spine, and certain general diseases, may in some instances produce impotency.

Complete sexual **incapacity or impotence in the female** is very rare. The most common congenital cause, perhaps, of such condition is the entire, or almost entire, absence of the vagina. In one reported case the vagina was contracted in depth, admitting the penis to perhaps less than half the usual extent, and ending in a cul-de-sac, without communicating with any of the internal organs; there appeared to be an entire absence of the uterus. The defect being deemed incurable and not admitting of complete copulation, the marriage was set aside.

In some cases an attempt has been made to form an artificial vagina by operation, where none exists, but such attempts can hardly be said to be successful, and doubtless the woman could not be called upon to submit to such an operation. Mere occlusion of the vagina by adhesion of the labia, or by an imperforate hymen may, however, in most instances be remedied by a slight operation; and if the woman should decline this operation, according to the rules already stated, a decree should be granted. A decree has been granted in several instances where there has been so very great a degree of sensibility of the vagina, accompanied in some instances by spasms, or perhaps by hysteria, as to amount practically to an impossibility of having sexual connection. Upon the principles already stated, even though such defects might be remedied by surgical or medical treatment, if the respondent should refuse to submit to such treatment a decree should be granted.

This subject and the kindred subject of sterility will be further considered in the chapter on LEGITIMACY AND PATERNITY. See, also, the next chapter.

CHAPTER XIV.

SEX, HERMAPHRODISM, AND MONSTROSITIES.

Sex. — The question of the sex of an individual is not unfrequently one of considerable importance, both as a matter of identification of the individual (which will be more fully considered in a subsequent chapter), and also in determining the rights of property depending upon courtesy, descent, etc.

If a body or considerable portions of a body are found, there will be little difficulty in determining the sex if the genital organs are present, except in rare instances of doubtful sex.

Besides the general contour of the shoulders and hips noticed below, the breasts of the female are, as a rule, more developed than those of the male. The pubic hairs of the male extend higher towards the umbilicus than those of the female; the distance between the navel and the pubes in the male is shorter than between the navel and the *scrobiculus cordis*; in females the reverse is the case. Males have more hair on the body, but less and shorter hair on the head. In the male the *pomum Adami* and the larynx are larger than in the female. The average male head is larger and the brain heavier than in the case of females.

Where a body is found in a state of decomposition, it should be remembered that the uterus has remarkable powers of resisting putrefaction, and the sex may often be determined from it when recognition from other organs has been rendered impossible by decomposition.

It should be remembered that until the age of puberty there is little difference in the general characteristics of isolated bones or complete skeletons of the two sexes, although generally male children are somewhat taller and heavier than females of the same age.

The general characteristics of skeletons or of the individual bones of adults, are as follows:—

In the male, the shoulders are broader than the hips; the bones usually present rougher and more prominent markings at the points of attachment of the muscles than those of the female, and the male skeleton generally exceeds that of the female both in height and weight.

In the female, the hips are broader, thighs shorter and larger, and the tuberosities of the ischia and the acetabula farther apart. The female skull is said to be smaller, more ovoid, more bulging at the sides, and larger behind the *foramen magnum* than that of the male. The face more oval, frontal sinuses less strongly marked, nostrils more delicate, jaws and teeth smaller, and the chin less prominent. The chest of the female is said to be deeper than that of the male; the sternum shorter and more convex; the ensiform cartilage thinner and ossified later in life; ribs smaller and the cartilages longer. The vertebral column is longer and the bodies of the vertebræ deeper in the female than in the male. The bones of both upper and lower extremities are generally smaller and lighter in the woman than in the man.

In many instances, however, it would seem to be impossible to determine the sex of an individual from an inspection of the skeleton without an examination of the pelvis. **The male pelvis** presents a narrow but deep excavation with smaller apertures; its bones are thick, its muscular impressions well marked and its angles abrupt and prominent.

The female pelvis is not so deep as that of the male but considerably exceeds it in its transverse and antero-posterior dimensions. The cavity is more capacious, its apertures larger, its walls less massive and rough, while its general contour is less angular and abrupt. The *alæ* of the *ossa innominata* spread further outwards, the anterior superior spinous processes and the tuberosities of the ischia and the acetabula being removed to a greater distance from the median line, whence arises the prominence of the hips in women. The sacrum is wider and less curved, and the sacro-vertebral angle less prominent than in the male. The *obturator foramen* is somewhat triangular in form and smaller in size than the male; the ischiatic spines project less into the cavity of the pelvis; the *coccyx* is more movable and the *symphysis pubis* less deep; the upper aperture is more nearly circular, its margin smoother and more rounded, the pubic arch is wider and more curved, and the rami are everted so as to present shelving surfaces rather than angular edges to any object descending through the perineal strait. The angle beneath the arch of the pubis is more obtuse than in the male. In the male this angle varies from 75 to 80 degrees, while in woman it varies from 90 to 100 degrees.

The above peculiarities of form are such as to permit the expansion of the uterus and the passage of the child in parturition.

The dimensions of male and female pelves according to Dr. Tidy are as follows:—

	Male.		Female.			
	Inches.	Lines.	Inches.	Lines.	Inches.	Lines.
Between the antero-superior spinous processes of the ilia	7	8	8	6 to	10	0
Between the middle points of the cristæ of the ilia	8	3	9	4 "	11	1
The transverse) diameter of the abdom-)	4	6	5	0 "	5	6
" oblique } inal strait of the true }	4	5	4	5 "	5	5
" antero-posterior) pelvis)	4	0	4	0 "	4	4
" transverse) diameter of the cavity)	4	0	4	7 "	4	8
" oblique } of the true pelvis }	5	0	5	2 "	5	4
" antero-posterior))	5	0	4	7 "	4	8
" transverse) diameter of the perineal)	3	0	4	0 "	4	5
" antero-posterior) strait of the true pelvis)	3	3	4	4 "	5	0

The diameters of the female pelvis are, according to Dr. Lusk, as follows:—

The diameters of Superior Strait or Brim:—

Antero-Posterior or (anatomical) conjugate, $4\frac{1}{4}$ inches.

The obstetrical conjugate diameter, measured about $\frac{2}{3}$ of an inch below the upper border of the symphysis, is only 4 inches.

The transverse or bis-iliac diameter of the brim, $5\frac{1}{4}$ inches.

Oblique diameter from the ilio-pectineal eminence to the opposite sacro-iliac articulation, 5 inches.

The circumference of the brim is very nearly 16 inches.

The diameters of the Inferior Strait or outlet of the pelvis: —

Antero-Posterior, 3¾ inches. When the coccyx is pushed back it may extend to 4½ inches.

Transverse diameter, 4¼ inches.

Owing to the elasticity of the sciatic ligaments, the oblique diameters are not of obstetrical importance.

The circumference of the Inferior Strait is 13½ inches.

Below the brim the dimensions are increased considerably by the concavity of the sacrum; thus a plane passing through the lower portion of the symphysis pubis and across the upper margins of the acetabula to the junction of the second and third sacral vertebræ, gains ¾ inches, in the conjugate, while the transverse diameter is barely ¼ inch less than the transverse diameter of the brim. The narrowing of the outlet is most marked in a plane drawn so as to intersect the spines of the ischia and the extremity of the sacrum; at this level the transverse diameter between the spines is but 4 inches, and the antero-posterior diameter 4½ inches.

Hermaphrodism. — The determination of sex in cases where the organs of generation are malformed is sometimes a matter of some difficulty, and occasionally nothing short of actual dissection after death has discovered the true sex of the individual. A full discussion of this subject is beyond the scope of this manual. The malformation is seldom confined to the genitals alone, but often extends to the conformation of the body generally, and even to the distinctive mental faculties, while the sexual desires are either wanting altogether or sometimes point in the wrong direction. In this connection only the more important facts will be stated.

Hermaphrodism may be divided into false and true. In the first species, the genital organs and general sexual conformation of one sex approach from imperfect or abnormal development, those of the opposite sex. In the second or true hermaphrodism, there actually co-exists upon the body of the same individual more or fewer of the genital organs and distinctive sexual characters, both of the male and female.   Strictly, perhaps, hermaphrodism should be defined as the co-existence in a single individual of completely developed ovaries and testicles, or one at least of each gland, the genital gland being the only reliable test of sex.   But in this strict sense it is probable that no true human hermaphrodite has ever existed, or at least such a one as has been able to perform the functions of either sex independently and to effect self-impregnation.

With proper care and attention it is always easy, says Dr. Ogston, to discover the true sex in the different varieties of false hermaphrodism.   It is chiefly in regard to the so-called true hermaphrodite that mistakes have arisen, even in the hands of competent observers.

The table of Sir James Simpson as altered by Dr. Ogston, is here given in order to facilitate the understanding of this subject : —

I. Spurious.	1. In the female, Androgynæ.	(1.) From enlarged clitoris. (2.) From prolapsus uteri.	
	2. In the male, Gynandri.	(1.) From extroversio vesicæ. (2.) From adhesion of the penis to the scrotum. (3.) From hypospadia.	
II. True.	1. Lateral.	(1.) Testis on the right, and ovary on the left side. (2.) Testis on the left, and ovary on the right side.	
	2. Transverse.	(1.) External sexual organs female, internal male. (2.) External sexual organs male, internal female.	
	3. Vertical or double.	(1.) Ovaries, imperfect uterus, vesiculæ seminales, and rudimentary vasa deferentia. (2.) Testes, vasa deferentia, vesiculæ seminales, and imperfect uterus and its appendages. (3.) Ovaries, and testes co-existing on one or both sides.	

1. In most cases where females have been mistaken for males, the enlarged clitoris is the prominent feature; a consideration, however, of the anatomical distinctions between the clitoris and the penis, which may be found in any work on anatomy, is sufficient to distinguish the two organs.

The prolapsed uterus could only be mistaken for a penis by one ignorant of the anatomical distinctions between the two organs.

2. Cases of spurious hermaphrodism in the male arising from the obtrusion of the bladder, the adhesion of the inferior surface of the penis to the scrotum, and from hypospadiac cleft of the penis, as a rule present no difficulty to the careful examiner; where, however, the gland is imperforate, and the penis itself diminutive in size, — resembling an enlarged clitoris, with the ure-

thra opening in a fissure of some length further back towards the perinæum, — there may be more difficulty. In such cases the scrotum is usually found split into two lateral halves, resembling the *labia majora* ; in such a case should the testes not have descended from the abdomen the question would be even more difficult. There are recorded instances of this sort where such a hypospadian has married and cohabited with a husband for a long term of years without the true sex having been discovered. In such cases the true criterion of sex is, as we have already stated, the presence of the distinctive genital glands, the testicle or the ovary. The determination whether a body in what appears to be a labium is a testicle or not is, however, not always easy ; it may be impossible of determination without dissection.

It should be remembered in this connection, that a very feminine appearance may result from non-descent of the testes or even of a testis, especially if atrophy of the organ has taken place.

The so-called true hermaphrodism may be considered under the three divisions given in the table.

1. A peculiarity of the sexual organs in cases of lateral hermaphrodism is that in the same individual there is found on one side of the median line what is or resembles a testicle and its appendages, and on the other side an ovary and its appendages. In such cases the ovary is more frequently found on the left side. In these instances, also, along with the testicle on one side and the ovary on the other, there generally coexists a more or less perfectly formed uterus, while the external parts of generation and of the body generally differ in their sexual characters, in some instances being female,

in others male, and in others of an indeterminate type. In some cases of lateral hermaphrodism spermatozoa have been found in the seminal fluid, while in others a periodic menstrual discharge has been recorded, — such conditions indicating the prevailing sex. Dr. Tidy records a case where were not only spermatozoa discovered in the secretion of the testicle, but regular menstruation was said to have occurred from the age of ten, and in which there were double sexual instincts.

2. In so-called transverse hermaphrodism, the external organs may be male and the internal female, or the reverse. By the terms external and internal are not meant those placed superficially or more deeply in the body. Those parts of the genitals which are covered by the common integument or are lined by the mucous membrane, and thus communicating with the external air, are classed as external, — including the vagina and uterus. The testis, though more exposed from being placed outside the pelvis, is termed an internal organ.

In cases of this sort of true hermaphrodism nothing short of dissection after death can determine the sex. The existence in such cases of a body which might be a testicle in the labium is considered by Dr. Ogston as too slight evidence whereon to form a decision.

3. Cases of the third variety of hermaphrodism cannot be verified during life, and their real nature can only be determined by dissection after death. In certain rare cases, according to Dr. Tidy, ovaries are associated with both male and female passages; and in other very rare cases, testicles are similarly associated; while in a third class both ovaries and testicles coexist in the same connection. Two cases are recorded by

him in which periodic menstruation and a seminal secretion containing spermatozoa occurred.

Sexless Beings. — According to Dr. Tidy, individuals are occasionally found presenting precisely opposite characters from those of hermaphrodites; namely, beings that have the essential features of neither males nor females. Several cases are recorded by him in which the genital organs are said to be entirely wanting in children born alive.

The following conclusions are arrived at by Dr. Tidy:

1. Given the presence of a testicle or testicles, wherever placed, and of a single opening communicating with a bladder and not with a uterus, — especially if there be seminal emissions containing spermatozoa and an absence of periodic hemorrhages, — the individual in question is to be accounted as a male, and that independently of anatomical malformations, such as the presence of an imperforate penis, or its entire absence, or the existence of the feminine configuration or instincts.

2. Given the presence of an ovary or ovaries, especially if there are periodic hemorrhages, the individual in question is to be accounted as a female, and that independently of anatomical malformations, such as the existence of a penis-like body, or of male configuration and instincts.

3. Given the presence of glands that may be either ovaries or testicles, and the precise nature of which there is difficulty in deciding; or given the absence of both ovaries and testicles, together with, in either case, the absence both of seminal emissions and of periodic hemorrhages, — then the presence of a uterus and of a second opening below and distinct from the

opening to the bladder, should be sought for. If a uterus or a second opening of the nature described be found, the individual is to be accounted as a female, and that independently of anatomical malformations, or of male configuration and instincts. But if, on the contrary, there be no uterus and no second opening below and distinct from the opening to the bladder, then the male sex is strongly indicated.

4. When the anatomical conditions are so equally balanced that neither sex seems to prevail, the existence of periodic hemorrhages is to be regarded as strongly indicative of the sex being female ; while on the other hand, the existence of emissions are strongly indicative of the sex being male. In the latter case, however, if spermatozoa can be detected in such individual, the proof that the same is male is complete.

5. Sexual inclinations, habits and tastes, and general configuration of the body, should in all cases be considered. If they support the conclusions based on the principles above laid down, they may be regarded as valuable confirmatory evidence ; but if, on the contrary, they fail to conform or even appear at variance with such conclusions, they may then be entirely disregarded.

Monstrosities. — According to Blackstone, it is a very ancient rule of the law of England, that "a monster which hath not the shape of mankind, but in any part evidently bears the resemblance of the brute creation, hath no inheritable blood and cannot be heir to any land, albeit it be brought forth in marriage ; but although it hath deformity in any part of its body, yet if it hath human shape it may be heir."

12

In a case of this nature, Dr. Tidy well says that "the duty of the medical jurist would be best performed by describing with the greatest detailed accuracy in what respects the individual in question differs from the human, leaving the court to say whether it be without the shape of mankind or not."

If the individual in question has the shape of mankind and is therefore capable of inheriting, the determination of its sex may be important with respect to questions of inheritance, questions of divorce, business, etc. This question has, however, already been considered.

The scope of this manual forbids our entering into the subject of Teratology in detail; for full information upon this interesting subject, the student is referred to professed treatises thereon.

In determining whether the monster has the shape of mankind or not, it may be stated that congenital deformities arising from a deficiency of structure, with arrested or defective development of parts, may give rise to acardiac monsters (monsters having no heart); acephalous monsters (monsters without a head); monsters anencephalic (having no brain, — the forehead, cranial vault, and brain being entirely wanting); or monsters with structural deformities of various kinds, such as the absence of arms, nose, eyes, ears, etc.

Acardiac and acephalous monsters can, of course, have no existence independent of their mother, and are not, therefore, likely to become the subject of medico-legal inquiry. Anencephalic monsters may at times, however, have a *medulla oblongata* and *cerebellum* and may be born alive; there are recorded instances of such a child having lived from one and a half hours to two months.

Children having deformities resulting from structural deficiency, such as the absence of arms, nose, etc., but having the vital organs perfect, as well as children having congenital deformities arising from a redundancy of parts, are occasionally born and live to adult age. No medico-legal questions, however, will arise respecting such persons, as they have all the rights of persons without any structural defect. Occasionally double monsters are born, where the children are more or less distinct above and below, being united by a band of greater or less width extending from the thorax and abdomen of one child to the thorax and abdomen of the other; or where the union exists between the back and· pelvis or between the head and scalp. Another class of double monsters is where the union is deep and intimate and more or less complete, — as where the children are single above and double below; or double above and single below; or where the bodies of the two children are so connected that they form a single body with a head at each end.

As to the medico-legal aspects of monstrosities, the only cases likely to demand the attention of the medical jurist are those of doubtful sex or hermaphrodism already considered, and those of monsters not possessing the shape of mankind. As to the latter, it seems that no such human monster that has lived to adult age could be denied human shape; and we quite agree with Dr. Tidy in saying that "it would constitute an almost unrecorded case where the jurist would be justified in saying without reserve that a child born alive had not the shape of mankind, — implying as the phrase does, far more than mere deformity."

It is very clear that by the common law no degree of monstrosity in a child will warrant the destruction of its life either by a physician or friend.

CHAPTER XV.

A LEGITIMATE child, according to Blackstone, is he that is born in lawful wedlock or within a competent time afterwards ; and a bastard is one that is not only begotten but born out of lawful matrimony. According to the statute in the State of Illinois, an illegitimate child whose parents have intermarried, and whose father has acknowledged him or her as his child, shall be considered legitimate. By the civil and canon laws the intermarriage of the parents after the birth of a child, legitimates such child.

It seems that a child born after the death of its mother as, for example, by Cæsarean section, is legitimate although begotten before the intermarriage of the parents.

Every child born in wedlock is to be regarded as legitimate unless impossibility of access or impossibility of intercourse with the husband can be proved by competent witnesses other than the husband or wife. A posthumous child will be legitimate unless non-access, impotence, or sterility can be proved. According to Lord Coke, "if a man hath a wife and dieth, and within a very short time after the wife marrieth again, and within nine months hath a child, so as it may be the

child of one or the other, some have said in this case
the child may choose his father."

Medico-legal questions relating to disputed paternity
may arise as to posthumous children; as to children
born either shortly after marriage or after the prolonged
absence of the husband; or in cases where a second
child is born a short time after the birth of another
child; where a child is born soon after a second marriage following the death of the first husband or a
divorce from him in cases where the decree of nullity
is not on the ground of impotence, which has already
been considered; and lastly, in cases of supposititious
children, where the woman during the life-time of her
husband pretends to have given birth to a child and is
suspected of substituting for the purpose of fraud the
child of some other person. These questions may be
considered under the following heads : —

1. Impotence and Sterility.
2. Duration of Pregnancy.
3. Similarity or Likeness.
4. Supposititious children.

1. **Impotence and sterility.** — Fruitfulness results
when a spermatozoon from the seminal fluid of the male
meets and becomes incorporated with an ovum discharged from the ovary of the female. The essential
conditions of fruitfulness then, are the existence of an
ovum capable of impregnation and the presence of
active spermatozoa. Impregnation may result altogether independently of sexual intercourse.

By impotence is meant inability for sexual intercourse; thus a male having no penis or a female no
vagina would respectively be impotent. The subject

of impotence, so far as it relates to the subject of divorce, has been considered in another chapter.

By sterility is meant an inability to conceive and to procreate ; a male whose semen contains no spermatozoa, or a female without ovaries, would respectively be sterile. We have already seen in another chapter that mere sterility, whether in the man or woman, is not in itself sufficient ground for avoiding marriage unless it is also accompanied by an irremediable incapacity for sexual intercourse.

Impotence and sterility in the man may arise from the male's not having arrived at the age of puberty. By the English common law the age of consent to marriage is fixed at fourteen in the male and twelve in the female ; and it is generally held that by the common law a rape cannot be committed by a boy under fourteen years of age. Although puberty may be considered as commencing somewhere about the age of fourteen in both sexes, exceptions to this rule are not uncommon ; puberty is not unfrequently long deferred, and again, there are recorded instances where boys have arrived at the age of puberty very early. Dr. Tidy records one case where the boy arrived at puberty at the age of four and a half years.

So far as respects the question of impotence and sterility in cases other than criminal prosecutions for rape, if unequivocal signs of puberty exist, the boy should be regarded as virile, no matter how young ; and if, on the other hand, signs of puberty are absent, the medical jurist has no ground for affirming the possession of virile power, however great the age.

There is no age beyond which parties may not legally

marry; and although old age usually lessens the virile
power, it cannot be regarded as negativing sexual fruitfulness or capacity.   Old age does, however, unquestionably render the occurrence of paternity less probable.
In such cases, however, the real question is not so much
the age of the man as whether his semen contains
spermatozoa.   Dr. Tidy says that he has more than
once detected spermatozoa in the semen of men over
ninety, and Casper records a case where he found them
in a man aged ninety-six.

The subject of impotence and sterility arising from
accidental injuries or congenital defects or malformations of the sexual organs — such as hermaphrodism,
hypospadias, epispadias, etc. — have been considered in
another connection.   In all these cases, so far as respects the question of legitimacy and paternity, the
principal points for inquiry are, —

Firstly, whether there is a urethra capable of per-
mitting the passage of the seminal fluid, and whether it
is so placed that the opening may in copulation come
into contact with any part of the vagina.

Secondly, whether the individual has a testicle or
testicles.   It is certain that fruitful intercourse may
result in the case of monorchids, that is, individuals
having only one testicle, or where only one testicle has
descended into the scrotum.   As respects the very rare
cases of crypsorchids, that is, where neither testicle has
descended into the scrotum, it may be stated that they
are not necessarily impotent.   As respects the question
whether they are sterile, recorded cases seem to establish the fact that a retained testicle does not, as a rule,
secrete prolific semen, and hence that they are generally

sterile. Casper, however, and Professor Owen have recorded cases not only of the existence of spermatozoa in the semen but of the possession of undoubted virility.

The removal of both of the testicles by operation certainly causes sterility; but it seems to be settled that, probably owing to the retention of semen in the *vesiculæ seminales*, the individual may for a limited period be capable of impregnating an ovum; and if the secreting structure of the testicles is not entirely removed, it seems clear that spermatozoa may be formed.

Where the testicles are congenitally absent the person will, according to Dr. Tidy, invariably be found to be of a languid disposition, slenderly formed, beardless, with a shrill voice, undeveloped genitals, and an absence of sexual desires. Where the testicles have not descended the development has been found, in some cases, to be in all respects manly and complete, and in others to be of a more or less womanly character. Where the testicles have been removed by operation a scar will always be found; in these cases if the testicles are removed in infancy the result will be much the same as where they are congenitally absent. If the testicles are removed after puberty, as a general rule the masculine character is retained, although exceptional instances to the contrary are recorded.

Impotence and sterility of the male may be caused by a variety of diseases, such as cancerous, tuberculous, or syphilitic diseases of the testicles or penis; an obstruction of the excretory ducts from double epididymitis following gonorrhœa. The ejaculation of semen may also be prevented by urethral stricture and by other pathological conditions. Certain general diseases, such as paraplegia, may also cause sterility by preventing sexual intercourse, especially so if accompanied with atrophy of the testicles. There are certain recorded instances where mumps has caused sterility in both sexes. Lithotomy has been known to cause sterility in the male by interference with the ejaculatory ducts. Finally, any cause which decreases the bodily vigor probably decreases sexual power; but there are recorded instances in which coitus in ar advanced state of phthisis or heart disease has been followed by pregnancy, although the coitus took place only a few hours before death.

Masturbation may be a cause both of sterility and impotence.

Impotence and sterility may arise from the action of certain poisons, notably alcohol, opium, tobacco, lead, and gonorrhœal virus ; the last mechanically by producing stricture of the urethra or of the seminal ducts. As to the poisons above mentioned other than the gonorrhœal virus, in the majority of cases it is perhaps doubtful whether their effects have not been exaggerated.

Impotence and sterility in woman may arise from a variety of causes. Impotence such as will warrant a decree for a divorce is extremely rare ; this subject has, however, been considered in another chapter.

Sterility may arise from extremes of age. The generative power in woman is generally supposed to commence with the appearance, and end with the disappearance of menstruation. The menses, or catamenia, from the arrival of puberty to the menopause or cessation of menstruation occur, as a rule, at regular intervals of twenty-eight days from the beginning of one to the commencement of the next monthly hemorrhage. The fluid discharged consists chiefly of blood and coincides in point of time with the discharge of the ovum from the ripened Graafian follicles of the ovary; the blood discharged has been supposed to be deficient in fibrin and is generally changed in color from admixture with vaginal secretions. Epithelial scales from the vagina and uterus will, as a rule, be found in such discharge. This flow may last from a few hours to several days, from three to four and a half days being the average period. The amount discharged may vary from a few drops to many ounces; an average quantity being from four to six ounces. Although the average interval, as above stated, is twenty-eight days, the flow may be anticipated or retarded by a variety of causes, and in some instances the normal interval varies within considerable limits. In cold countries the interval is said to be longer than in warm countries. As a rule, menstruation is said to begin in temperate climates somewhere between the ages of fourteen and sixteen, — the usual earliest periods being twelve to thirteen, and the latest periods nineteen to twenty-three years, although exceptional instances of primary menstruation have been recorded as early as one year and as late as forty-seven; primary menstruation, however, before the age of nine and after twenty is exceedingly uncommon.

With female puberty come certain changes, — namely, deposit of fat in the subcutaneous cellular tissue, causing first a slight swelling in the groins, extending over the whole body, more particularly the breasts; growth

of hair in the arm-pits and upon the genitals, while the
hair of the head often assumes a darker appearance;
the voice changes from its childish quality to the
characteristic voice of woman.

The catamenia, as a rule, cease between forty-two and
forty-eight, although there are numerous exceptions.
As a rule, those who develop early fade early; and
there are recorded cases where, after the change of life
seemed complete, menstruation reappeared and continued to a late period. The fruitful period in woman is
generally limited to the interval between the first and
last menstruation, but it is an established fact that
menstruation may occur without ovulation. There are
recorded cases of so-called menstruation from birth, but
the earliest recorded case of pregnancy was between
eight and nine years.

As to whether ovulation may occur without menstruation, there is no doubt; recorded cases of the conception of women who have never menstruated furnish
important evidence upon this point. There are also
recorded cases of women conceiving who menstruated
regularly before marriage and irregularly or not at all
during their child-bearing life; and there are also cases
of conception after the menopause. As to what time
during the monthly period the ovum is discharged from
the ovary, there is much doubt; it may be regarded as
settled, however, that while pregnancy is more apt to
follow a few days after the cessation of the monthly flow,
or a day before its commencement, conception may take
place at any time during the interval. As to the age
beyond which conception is impossible, the limit doubtless varies with different women; and while, as a rule,

ability to conceive ceases with the menopause, there are, as we have stated, recorded cases occurring after that time. It is not possible, perhaps, for the medical jurist to fix upon any age before or after which he will be justified in stating that pregnancy is impossible. Pregnancies up to fifty are not very uncommon; between fifty and sixty, although there are some well authenticated instances, the cases are very uncommon, while there are recorded cases above the age of sixty, and even as late as seventy; these cases, however, do not seem entitled to much credit.

Impotence and sterility in women may arise from congenital and other defects, such as the absence of the vulva and external genital organs; absence of the vagina; absence of a uterus, which may or may not coexist with an absence of the vagina; or entire absence of ovaries, which, however, could not be demonstrated during life.

Sterility may be caused by the non-performance of their normal functions by the ovaries, or by some interference with the passage of the ovum from the ovary to the uterus, arising from alteration in the coverings of the ovary, in the Fallopian tubes, etc.; it may arise from alterations in the uterus or vagina which prevent the impregnation of the ovum or access to it by the spermatozoa; or from the destructive action of unhealthy fluids of the uterus or the vagina upon the spermatozoa. Diseases and displacements of the uterus, contractions of the *os uteri* and the cervical canal, tumors in the uterine cavity, etc., may also cause sterility.

As a general rule, women do not conceive during lactation, but the exceptions to this rule are not infrequent. Promiscuous intercourse, as by prostitutes,

from the inflammatory condition of the Fallopian tubes
thereby induced, frequently renders impregnation less
likely to occur. The habit of abortion at an early period
of pregnancy often practically amounts to sterility.

The full consideration of this subject, however, is
beyond the scope of this manual; the student is referred
for further particulars to professed treatises on obstetrics
and gynæcology.

A microscopical examination of the semen to ascertain whether it contains spermatozoa, and if it does
contain them whether they are alive, is often an important aid in determining the question of sterility.

2. **The duration of pregnancy, etc.,** has an important
bearing upon questions of legitimacy and paternity.
The signs of pregnancy, time of quickening, etc., have
already been considered in another connection. As to
its duration, it is impossible to fix definitely a period of
utero-gestation. Although some foreign codes of jurisprudence have attempted to fix the time beyond which
utero-gestation cannot extend, the common law has more
properly left each case to be decided according to the
evidence introduced; forty weeks, however, or two hundred and eighty days, is commonly regarded as the
period of pregnancy.

Dr. Tidy, in the consideration of this subject, states
that there is a general consent among the best of obstetricians as to the duration of pregnancy, the extremes being 266 days, or 38 weeks, to 280 days, or 40
weeks.

Duncan, from analysis of 46 cases, in which connection took place during a single day only, found the
average time to the date of parturition to be 275

days. Ahfeld, from analysis of 425 cases, obtained an average of 271 days. Hecker, from 108 cases, found the average to be 273.5 days. Veit, from 43 cases, found an average of 276.4 days. Faye, from 63 cases, found an average of 270.7 days.

While these averages show that the date of a particular confinement will probably occur at the time indicated thereby, the fact that in Ahfeld's table there exists between the longest and shortest gestation a difference of 99 days, in Hecker's a difference of 63 days, and in Veit's a difference of 36 days, shows that too much reliance should not be placed upon such averages. The bulk of confinements, however, vary within narrow limits. According to Lusk, of 653 women, in 15.93 per cent delivery occurred in the 38th week; in 27.56 per cent, in the 39th week; in 26.19 per cent in the 40th week; in 10.01 per cent in the 41st week. Thus in more than half the cases delivery occurred in the 39th and 40th weeks, and 80 per cent occurred between the 38th and 41st weeks inclusive. Of the remainder, 14 per cent took place prior to the 38th week, and according to Dr. Lusk, were probably influenced by many operative accidental causes which favor prematurity. Of the 6 per cent reported as occurring later than the 41st week, a considerable number he regards as of questionable authenticity, as in his opinion gestation protracted beyond the 285th day is certainly of very rare occurrence.

The extent to which the normal period of utero-gestation may be shortened, and a living child nevertheless be born, is one of great importance. In questions of this sort, two points usually arise for consideration:

thetheththththtthththtororororororororororororororororororororororororororrororororororororIapologize—Ineedtoprovidetheactualtranscription.

first, whether a child of the age stated or estimated is viable, — that is, of such an age as to be capable of showing some indications of live birth after it is completely external to the mother; and secondly, whether, if it is alive when born, it is probable that, at the age stated or estimated, it can be reared.

Dr. Tidy draws the following practical conclusions from the consideration of a large number of cases of abnormally shortened utero-gestation: —

1. Allowing that from the first moment of impregnation the ovum is truly alive, and further that mere motion of limbs, or evidence of circulation without active respiration, are sufficient to constitute live birth, — nevertheless, there is no evidence to show that a fœtus born at an earlier period than between the fourth and fifth months of uterine existence can in any sense be said to be born alive, much less lead an independent life apart from its mother.

2. That living children have been born between the fourth and fifth months of uterine life. As a rule, however, the only sign of life exhibited by children born at this early period is a slight motion of the limbs, although cases of somewhat more active vitality have been recorded. There is, however, no well authenticated case where a less than five months' child has lived beyond twenty-four hours after its birth, and but one where it has lived twenty-four hours.

3. That children born alive at the fifth or between the fifth and sixth months of utero-gestation mostly die after a few hours. Nevertheless, there is a limited number of recorded cases where such children have been reared, and have even reached adult age.

4. That several well-authenticated cases exist where children born between the sixth and seventh months, and even at the sixth month, have reached adult age; but that in such cases more than ordinary care and attention have been needed to maintain life, at least for some time after birth.

Signs of Maturity in New-born Children. — Although children born at full term vary in size, weight, etc., there are certain signs of maturity which should be considered in this connection. Among these signs are,—

1. A certain general *habitus* familiar to experts.

2. *The color of the skin* of a mature child is paler than in one less mature, while the down (*lanugo*) to a great extent disappears with maturity. It is said that certain white points found in many cases on the *alœ* of the nose, cheeks, and forehead, but especially on the chin and under lip,—due to the dilatation of the excretory ducts of the sebaceous follicles, — are abundant in proportion to the immaturity of the fœtus, decreasing in number as full term approaches; at full term they are only to be found on the tip of the nose.

3. At maturity there is *more or less hair* on the head.

4. *The immobility of the skull bones,* — the anterior fontanelle averaging $\frac{3}{4}$ to 1 inch in length.

5. *A certain height and weight.* The average height of new-born children of both sexes, born at full term, is between 18 and 19 inches. Of 247 mature children of both sexes, the average length is stated by Dr. Tidy to have been $18\frac{4}{5}$ inches. Average length of the males (130) was $19\frac{5}{8}$ inches; of the females (117) $18\frac{5}{8}$. The maximum recorded length was 22 inches; the minimum,

16.   The average weight of a mature infant at birth is probably between 6 and 7 pounds. Dr. Tidy states that the average weight of English children at birth is 6 pounds, 8 ounces. The average weight of 247 mature children of both sexes is stated by him to be $7\frac{1}{20}$ pounds; of 130 males, $7\frac{1}{3}$; of 117 females, $6\frac{4}{5}$. The maximum recorded weight of these children was 10 pounds, and the minimum, $4\frac{1}{2}$ pounds. It is said that the weight and length of Indian children are less than those of European children.

Casper quotes from Gunz the following table of dimensions of the bones of a mature child:—

	Inches.	Lines.
Height of the frontal part of the frontal bone	2	3
Breadth of the same	1	10
Length of the *pars orbitalis*	1	
Breadth of the same	1	
Parietal bone from anterior superior angle to inferior posterior one	3	3
Ditto from anterior inferior angle to superior posterior one	3	3
Height of *pars occipitalis* of *os occipitis*	2	
Breadth of the same	1	10
Height of squamous portion of temporal bone from upper edge of auditory foramen.	1	
Height of malar bone		6
Breadth of the same	1	
Height of nasal bone		5
Breadth of the same		3
Height of the superior maxillary bone from the *processus alveolaris* to the apex of the *processus nasalis* . . .	1	
Length of the superior maxillary bone from the *anterior nasal spine* to the apex of the *processus zygomaticus* .	1	1
Length of each half of the lower jaw	1	10
Breadth of the lower jaw		7
Length of the seven cervical vertebræ	1	3
" " twelve dorsal "	3	9

	Inches.	Lines.
Length of the five lumbar vertebræ	2	3
" " sacrum and coccyx	2	3
" " collar bone	1	7
" " shoulder blade	1	6
Breadth of the shoulder blade	1	2
Length of the humerus	3	
" " ulna	2	10
" " radius	2	8
" " femur	3	6
" " patella		9
Breadth of the patella		8
Length of the tibia	3	2
" " fibula	3	1

It is laid down as a general rule that still-born children are heavier and longer than those born alive; males than females, and single children and twins than triplets.

6. **The diameter of the head** and measurements across the shoulders and hips, according to Casper, afford valuable evidence of maturity. Of 207 children he found the average transverse diameter of the head $3\frac{1}{5}$ inches; longitudinal diameter, $4\frac{1}{8}$ inches; diagonal diameter, $4\frac{7}{8}$ inches. Tardieu makes the occipito-frontal diameter from $4\frac{1}{3}$ to $4\frac{1}{2}$ inches, and the biparietal from $3\frac{5}{8}$ to $3\frac{3}{4}$.

According to Lusk, whose table is based upon that of Tarnier and Chantreuil, the diameters of the fœtal head are as follows :—

Occipito-mental diameter	$5\frac{1}{4}$ inches.
Occipito-frontal "	$4\frac{1}{2}$ "
Sub-occipito-bregmatic	$3\frac{3}{4}$ "
Bi-parietal	$3\frac{3}{4}$ "
Bi-temporal	$3\frac{1}{4}$ "
Bi-mastoid	3 "
Fronto-mental	$3\frac{1}{4}$ "
Cervico-bregmatic	$3\frac{3}{4}$ "

According to Casper the average diameter across the shoulders in 117 mature children was $4\frac{15}{16}$, and across the hips, $3\frac{6}{13}$ inches.

7. **Condition of the nails.** — In mature children the nails are horny and reach the tips of the fingers, although not necessarily so in the case of the toes.

8. **The cartilages of the ears and nose** of a mature child feel cartilaginous.

9. **Condition of the genitals.** — In mature male children the testicles will both be in the scrotum, which will be corrugated. In about three or four per cent of cases, according to Dr. Tidy, one of the testicles will not have descended at birth; it is rare to find both testicles undescended. In mature female children the labia majora should cover the vagina and clitoris; there are, however, numerous exceptions to this rule.

10. **Position of the Umbilicus.** — The navel is usually in the centre of the body of a mature child; there are, however, numerous exceptions to this rule.

11. **Weight of the placenta and length of the cord.** — The placenta at full term usually has a diameter from 8 to 10 inches, and weighs from 15 to 19 ounces. The length of the cord at full term is stated by Tidy to vary from 18 to 21 inches.

12. **Ossification of bones, etc.** — Where the child is born dead the following sign is of value: In the second half of the tenth lunar month, the centre of ossification of the inferior femoral epiphysis makes its appearance; this centre of ossification is to be demonstrated by carefully making horizontal sections off from the cartilaginous epiphysis until a colored point is observed, the greatest diameter of which osseus nucleus is

to be measured.   To the naked eye this nucleus appears as a more or less circular blood-red spot in the midst of milk-white cartilage, in which vascular convolutions may be distinctly recognized.

If there is no visible trace of this osseous nucleus, the fœtus cannot be more than from 36 to 37 weeks old; if it is of the size of a hemp-seed (about half a line), it corresponds to 37 or 38 weeks, if still-born; when from 3-4 to 3 lines it indicates a uterine age of about 40 weeks; if it measures more than 3 lines the child has probably survived its birth.

The following table quoted by Tidy from Tardieu, shows the character of the fœtus at different ages of intra-uterine development.

Table showing the Character of the Fœtus at different Ages of Intra-Uterine Life.

Age.	General Development of the Body.		State of the Skin, etc.	Degree to which Ossification has advanced.
	Height.	Weight.		
From 1 month to 1½ months. From 1½ to 2 months. From the 2d to the 3d month.	½–¾ inch. ¾–2 inches. 2–4 "	15–46 grs. 77–155 " 3 vj–1½ oz.	Skin quite transparent, of a purplish-red color, with no trace of hair on it.	Centres of ossification for clavicles and lower jaw. Appearance of dental papillæ in the furrow of the lower jaw.
From the 3d to the 4th month.	4–6 "	1½ oz.–4½ oz.	Development of nails. Appearance of matrix of nail. Sex distinct.	Centre of ossification in *ischium*.
From the 4th to the 5th month.	6–8 "	6–8 oz.	Hair-germs appear on the forehead and eyebrows.	Ossification of *os calcis*.
From the 5th to the 6th month.	10–12 "	8–12 oz.	Hairs appear on the limbs.	Osseous centres for *astragalus* and *os pubis*.
From the 6th to the 7th month.	12–14 "	15–32 oz.	Hairs on hands and feet. *Membrana pupillaris* begins to disappear.	Three or four osseous centres in *sternum*.
From the 7th to the 8th month.	14–16 "	2–3 pounds.	Skin has lost its transparency. Epidermis distinct. Color pale pink.	Ossification of lower vertebræ of *sacrum*.
From the 8th to the 9th month.	16–18 "	3–5 "	Skin is covered with sebaceous materials. (*Vernix caseosa*.) The nails do not reach tips of fingers.	
At term (mature).	18–20 "	6–7 "	Sebaceous covering still thicker. Nails overlap fingers. *Membrana pupillaris* has quite disappeared. Navel a little below middle of entire length of body.	An osseous nucleus in the condyloid epiphysis of *femur*. The alveolar processes of the lower jaw are perfectly distinct.

The student will find an extended consideration of the evidence of the development of the fœtus in 2 "Tidy's Legal Medicine," pages 59 to 64 inclusive.

13. The extent to which the normal period of utero-gestation may be lengthened, in practice must be reckoned from the latest possible day of access of the husband, which in most cases would be the date of his death, departure abroad, or the like. That the period of utero-gestation may be considerably prolonged beyond the average period, there can be no doubt; but, as has already been stated, by the law in some countries a definite period is fixed beyond which the child born is regarded as illegitimate. Thus in Scotland a child born six months after the marriage of the mother or within ten months after the death of the father, is considered legitimate. In France the limit is fixed at 300 days ; the Prussian code, without absolutely declaring children born in the 11th month illegitimate, attaches such conditions to the proof of their legitimacy as to make it almost unattainable. By the English common law the question is always one of fact, depending upon the evidence.

Obstetricians have furnished us with numerous instances of utero-gestation extending beyond the 40th week, as calculated from the cessation of the catamenia. Dr. Ogston has tabulated 155 such cases, of which —

55	were protracted	to	the 41st	week	or to	287	days.
42	"	"	42d	"	"	294	"
30	"	"	43d	"	"	301	"
13	"	"	44th	"	"	308	"
12	"	"	45th	"	"	315	"
3	"	"	47th	"	"	322–325	"

He also tabulates 55 cases dating from a single coitus, as follows : —

2 cases protracted to	281 days, or	40 weeks,	1 day.
2 " "	283 "	40 "	3 "
2 " "	284 "	40 "	4 "
2 " "	286 "	40 "	6 "
4 " "	287 "	41 "	
1 " "	288 "	41 "	1 " .
1 " "	289 "	41 "	2 "
1 " "	291 "	41 "	4 "
3 " "	293 "	41 "	6 "

or 8 were protracted into the 41st week, 7 into the 42d, while 3 had almost reached the 43d week. From these cases he arrives at the conclusion that utero-gestation may in a few instances be extended to the 44th or even to the 46th week.

Upon the same subject Dr. Tidy concludes that there is considerable evidence to show that 41, 42, and 43 weeks may elapse between coitus and labor; and that although it is impossible to state that 44 weeks or even longer periods of gestation may not occur, it must be conceded that there are no well-recorded cases of such protraction, except where the time has been determined by the cessation of the catamenia.

In cases of this sort, moral considerations, such as the character of the parents, probability or possibility of access, are matters of great importance.

In cases where pregnancy is much prolonged beyond the usual period, as well as in doubtful cases, the possibility of the more or less complete development of the ovum in the ovary, in the Fallopian tube, in the walls of the uterus, or in the peritoneal cavity, must not be lost sight of. In the peritoneal cavity the ovum may proceed full development, become encysted, and remain dormant for years; but in the other cases mentioned, pregnancy will not be likely to proceed farther than the first half of the usual term without causing the death either of the fœtus or of the mother.

In this connection, the possibility of *super-fœtation* or the impregnation of a second ovum in a woman already pregnant, should not be lost sight of. Many of the supposed cases of super-fœtation may doubtless be explained as being twin or triplet births, where a considerable interval has elapsed before the birth of each child after the first. A case of triplets is recorded where there was an interval of fifteen days between the successive births. It should also be remembered that there are instances of fœtuses remaining *in utero* for a considerable time after their death, even when there is no other fœtus alive in the womb; and that at the fifth and eighth month of pregnancy dead fœtuses of not greater development than from three to five months have been born; and that not unfrequently dead fœtuses of different degrees of development have been retained until the expiration of full term. Excluding, however, all cases which are liable to be mistaken as cases of super-fœtation, there are a few cases which are difficult of explanation on any other hypothesis than that a second impregnation must have taken place while a partially developed fœtus was in the womb.

In determining whether the case is one of super-fœtation or not, the following points should be investigated: the size and development of each child, and the interval between the births. If the children are both mature, and the interval between the births is two months, super-fœtation is probable, since it is altogether improbable that a seven months' child should possess the

appearances of maturity of a nine months' child. If the
interval varies between two and three months, and the
first child is immature and the second mature, the case
is more likely one of twins, where one is born prema-
turely and the other at full term. Where, however,
there is an interval of four months between the births,
and both children are capable of being reared, super-
fœtation may be admitted, as it is highly improbable
that a five months' child should be capable of being
reared.

3. **Similarity, or likeness.** — Most authors upon medical jurisprudence regard likeness existing between the
child and its alleged father, as strong evidence of paternity; and even Lord Mansfield is reported to have said,
" I have always considered likeness as an argument of a
child being the son of a parent." The absence of likeness is certainly not evidence of non-relationship, and
there is great difficulty in drawing any definite conclusion from mere likeness. It has been definitely held in
a case recently decided by the supreme court of Wisconsin (*Hanawalt* vs. *The State*, 64 Wis. 84), that in a
bastardy case the child may not be exhibited to the jury
as evidence of paternity.

In questions of this sort the possibility of atavism,
or resemblance of a child to some ancestor prior to
its father or mother, should be remembered; also that
in cases of a second marriage by the mother, a child of
the second marriage may resemble neither parent, but
the first husband.

In questions of paternity considerable importance is
properly attached to the transmission of color; and
there are instances where the paternity of the father

who was a black man, has been established by the color of the child. The case of the *Commissioners* vs. *Whistelo* (3 Wheeler Crim. Cases, 194) is an interesting case in this connection.

4. **Supposititious children.** — Cases of pretended delivery or alleged substitution of one child for another should be examined by the medical jurist upon the principles stated in a preceding chapter. Without the least delay, careful examination of the person of the woman should be made in order to ascertain whether there are or are not signs of recent delivery. A demand should be made to see the placenta, and the child should be examined with reference to ascertaining whether its appearance agrees with that of the alleged mother. Stains upon the bedding and marks of blood may be fraudulently made; and there are instances of the substitution of some viscus of an animal for the placenta.

CHAPTER XVI.

PERSONAL IDENTITY.

General Considerations. — Questions of personal identity often become matters of inquiry in the course of judicial proceedings. Such cases will usually turn upon the evidence of ordinary witnesses without the aid of the medical expert; not unfrequently, however, such expert is able to furnish important aid in determining such questions.

The question may arise in the case of a living person or of a body which has been dead only a short time; with respect to bodies long dead or with respect to complete skeletons or detached portions of the same. In the case of a living person or of a body only recently dead, the sex, occupation, complexion, general type of face, race, hair, nails, stature, scars, injuries, congenital malformations, as well as the clothes, etc., upon the body, form important matters for investigation. In cases where the remains are mutilated or only a portion of the body is discovered, the question usually presents more difficulty. The case of *The Commonwealth* vs. *Dr. Webster*, for the murder of Dr. Parkman, reported in 5 Cush. 295, 386, is an interesting one in this connection. See also Bemis's report of this case. In such cases the stature may often be approximately determined from the measurement of certain bones and their known relation to the height of the person. The correspondence or otherwise of the several parts will sometimes enable one to determine the question whether they are parts of the same body or not. The method of mutilation is sometimes suggestive. The cause of the death may in some instances be determined by the examination of mutilated remains.

The extent of the destruction of the soft parts, as bearing upon the time of death, is often important to be considered. Clothes, buttons, jewelry, or other articles surrounding the body, if there are any, should be carefully examined and preserved.

Bodies long dead, skeletons and detached bones. — The length of time required in an ordinary grave to reduce a body to a skeleton is supposed to be about ten years; certain parts, such as the hair and nails, will resist decomposition much longer, if not indefinitely. The subject of decomposition of a body has, however, been fully considered in another place. The position and relation of the bodies to their surroundings may be of aid in determining their character and identity. In Christian burial it is customary to place the body at full length in the ground, with the head towards the west. Remains of clothes, buttons, jewelry, etc., as already stated, will often afford valuable aid in the identification of a body. Where the whole skeleton is submitted for examination there can be no difficulty in deciding whether the bones are human or not; and in ordinary cases there ought not to be much difficulty in determining whether single bones preserved entire, are or are not human. It would be presumptuous, however, in most instances where the bones are not considered human to undertake to determine to what animal they belong.

By the aid of the microscope it is possible to determine whether the smallest fragment found is or is not bone ; but according to the present state of our knowledge it is not possible thus to determine whether a particular fragment is part of a human bone or of some other animal.

It sometimes becomes important to determine whether bones are parts of the same skeleton or belong to more than one body; this can only be determined by comparison and placing each bone in the place where it belongs, aided by a competent knowledge of human anatomy. Where the bones are fractured or otherwise injured, it is important to determine whether the injury was caused during the exhumation or during life ; and if during life, whether or not it was of recent occurrence. If new bone is found around the broken ends this is proof that the fracture occurred some time before death. Diseased conditions of a bone may be discovered, and in such case should carefully be distinguished from decay and violence. All personal deformities should be carefully noticed. The presence of fœtal bones about the skeleton of a female, suggests but does not prove pregnancy.

The age may sometimes be approximately determined by inspection of the bones, a subject, however, which will be further considered in the next section.

Age.—The determination of age during life is usually a matter of considerable difficulty. In the case of children it may usually be determined approximately by observing the teeth, height, weight, and general development; in cases of hereditary syphilis and other diseased conditions, the development may be retarded.

During the middle period of life there is very great difficulty, if not an impossibility in determining the age; as the person becomes older this difficulty diminishes somewhat. In females of advancing age the mammæ either waste or enlarge considerably; in old people the arteries often become cord-like and tortuous, and the *arcus senilis* appears in the eye. The figure stoops as the intervertebral substance is absorbed, and the muscular power lessens; the teeth decay or come out, and the person puts on the characteristic appearance of senility, which being familiar to every one needs no further description. In determining age from the skeleton, important evidence is often derived from the points of ossification and the extent to which osseous union has progressed. The following details under this head are taken from Dr. Tidy's work on Legal Medicine:—

1 *year:* Points of Ossification, — lower extremities of humerus and ulna; heads of the femur and humerus; upper cartilage of tibia.

1½ *years:* Anterior fontanelles should be closed.

2 *years:* Points of Ossification, — lower cartilages of radius, tibia, and fibula.

2½ *years:* Points of Ossification, —greater tuberosity of the head of the humerus; patella; lower ends of the last four metacarpal bones.

3 *years:* Points of Ossification, — the trochanters.

4 years: Points of Ossification, — the second and third cuneiform bones of the tarsus.

4½ years: Points of Ossification, — the small tuberosity of the head of the humerus ; the upper cartilage of the fibula.

6 years: The descending ramus of the pubis meets the ascending ramus of the ischium.

From 8 *to* 10 *years:* The upper cartilage of the radius becomes ossified.

9 years: The ilium, ischium, and pubis meet in the cotyloid cavity (acetabulum) to form the pelvis.

10 years: Ossification begins in the cartilaginous end of the olecranon.

12 years: Points of Ossification, — the pisiform bones of the carpus.

13 years: The three portions of the os innominata (ilium, ischium, and pubis), though nearly united, can still be separated. The neck of the femur is ossified.

14 years, or about puberty : There are now added some fourteen additional centres to the sacrum.

15 years: The coracoid process becomes united to the scapula.

Between 15 *and* 16 *years:* The olecranon becomes united to the ulna.

From 18 *to* 20 *years:* The epiphysis at the upper end of the thigh bone is joined to the body of the bone, as well as those belonging to the metacarpus, metatarsus, and phalanges.

20 years: The upper and lower epiphyses of the fibula, as well as the lower epiphysis of the femur, are respectively united to the bones.

25 years: The epiphysis of the sternal end of the clavicle, and of the crista ilii are united to the bones.

If all the epiphyses be found united to their bones, and the bones themselves are solid and well-marked as to muscles, processes, and foramina ; and further, if the jaws show the wisdom teeth, — we may conclude the individual to be of adult age.

The Vertebræ. — The epiphyses of the bodies of the vertebræ are sometimes not consolidated until thirty years of age.

The Cartilages of the Larynx in advanced life assume more the appearance of osseous than of cartilaginous structure.

The Sternum. — The second and third pieces of the sternum rarely join until the thirty-fifth or fortieth year, while the union of the first and second pieces is not usually complete until quite advanced life.

The Cartilages of the Ribs. — The cartilages of the ribs generally ossify late in life. Dr. Humphrey regards this ossification as rather a sign of disease than of age. The first cartilage is more frequently ossified, and at an earlier period of life, in men than in women.

The Skull. — In old age the diploë is more or less absorbed, leaving the cranial bones thinner than they were in middle life. The sutures become firmly ossified, and gradually less distinct. If the sutures of the skull are indistinct, we may then fix the age as at least between fifty and sixty. As a rule the parietal sutures disappear about the age of puberty, although sometimes (but rarely) they remain separate throughout life.

The Lower Jaw. — The alveolar cavities containing the teeth are formed about the sixth month of intra-uterine life, while the rudiments of the whole of the temporary and of some of the permanent teeth (the anterior molars, for instance) are usually found within the gums in capsules at the time of birth. Again, the

14

jaw of the infant is rounded and somewhat semi-circular, the ramus and body forming an obtuse angle. As age advances towards middle life, the jaw loses its roundness, and as the alveolar processes containing the teeth become more and more perfect, exhibits a well-marked angularity and squareness. With old age, the teeth drop out or decay away, the jaw returning to its infantile shape. After a time the whole alveolar body may become absorbed, a sharp ridge replacing the holes for the teeth as they originally existed.

The Femur. — The neck of the femur *before puberty* is directed obliquely, so as to form a gentle curve from the axis of the shaft. In the *adult* male it forms an obtuse angle with the shaft, being directed upwards, inwards, and a little forwards. In the *female* it approaches more nearly to a right angle. Occasionally in very *old* subjects, and more especially in those greatly debilitated, its direction becomes horizontal, so that the head sinks below the level of the trochanter, and the length diminishes to such a degree that the head becomes almost continuous with the shaft.

The subject of *sex* has already been considered in another chapter.

Stature and Weight. — The average lengths of the fœtus at different periods of intra-uterine life, and of children and adults at various ages, are variously stated by different authors.

The following table is taken from M. Sue : —

Age.	Total Length.	Trunk.	Upper Extremities.	Lower Extremities.
Fœtus of 6 weeks.	16 lines.	1 inch.	5 lines.	4 lines.
2¼ months.	2 in. 3 lines.	1 in. 8 lin.	9 lines.	7 lines.
3 "	3 inches.	2 in. 1 "	13 lines.	11 lines.
4 "	4 inches 4½ lines.	2 in. 11 "	1 in. 9 lin.	1 in. 5½ lines.
5 "	6½ inches.	4 in. 4 "	2 in. 6 "	2 in. 2 "
6 "	9 inches.	5 in. 8 "	3 in. 7 "	3 in. 4 "
7 "	1 foot some lines.	6 in. 5½ "	5 in. 10 "	5 in. 9 "
8 "	14 inches 0 lines.	8 in. 3½ "	6 in. 8 "	6 in. 6 "
9 "	18 inches.	10 inches.	8 inches.	8 inches.
1 year.	22½ inches.	13 in. 6 lin.	9 inches.	9 inches.
3 years.	2 ft. 9 in. some lin.	19 inches.	14 inches.	14 in. some lin.
10 "	3 feet 8¼ inches.	2 feet.	19 inches.	20 in. 6 lines.
14 "	4 feet 7 inches.	2 feet 4 in.	24 in. 6 lin.	27 inches.
25 "	5 feet 4 inches.	2 feet 8 in.	30 inches.	32 inches.

Healthy children of healthy mothers, as a rule, diminish considerably in weight soon after birth, such decrease amounting to about 1-14th of the total weight at birth. In about 60%, children begin to increase in weight about the fourth day, and about the tenth day the weight will be the same or slightly in excess of that at birth. The increase in a child's weight thereafter should be progressive; much depends, however, upon the hygienic conditions surrounding it and the nature and amount of its food, etc.

The following table, taken from Tidy, gives the average monthly weight of young children during the first year:

	lbs.	oz.			lbs.	oz.
At birth	6	8	7 months		13	4
1 month	7	4	8 "		14	4
2 months	8	4	9 "		15	8
3 "	9	6	10 "		16	8
4 "	10	8	11 "		17	8
5 "	11	8	12 "		18	8
6 "	12	4				

The average weight of male adults at 20 is 143 pounds, and of females, 120 pounds. (See *post*, INSUR-ANCE.) As a rule, women increase in weight up to nearly fifty; while males so increase only up to thirty-five; in advanced age both sexes weigh about fifteen times their weight at birth.

The following table, from Barthes & Rilliet, is founded on the measurements of thirty-seven well-nourished and well-grown children: —

	3½ to 5 years.	6 to 10 years.	11 to 15 years.
Height	Inches. 32.8–38	Inches. 38–50.8	Inches. 50–52.8
Length of Sternum	4.4–5.2	4.8–6	5–7.2
Length of Dorsal Vertebræ .	5.6–8.8	7.2–10.4	9.2–11.6
Space between the Coracoid processes	5.2–6.8	6–8	7.6–10.8
Round the Thorax (under Armpits) during inspiration while seated	20.4–24.4	22.4–26.8	27.6–35.2
Round the Thorax (under Nipples) during inspiration . .	22–25	22.4–26.8	27.2–32

The average height of the adult French male is said to be 63 inches, and of the adult French woman, about five feet. In England the average height of men born under favorable circumstances, is said by Dr. Tidy to be 5 feet, 9 3-5 inches, and of women, 5 feet, 2 inches. The head is ⅛ part of the total height, and is divided into two equal parts immediately below the eyes, the nostrils being midway between the eyes and chin. The pubis is the central point between the two extremities of the body. When the arms are raised vertically above the head, the navel is the centre of the length; if the arm is divided into five parts, the hand occupies one part and the fore-arm and arm two parts each; multiply, therefore, the length of the hand by five and we get the length of the arm. The carpal and metacarpal bones represent half the length of the hand; the first phalanx of the middle finger is $\frac{1}{4}$ the length of the entire hand, its last two phalanges being together equal to the length of the first; the last phalanx is halved by the nail. The sole of the foot is $\frac{1}{3}$ longer than the palm of the hand, but the back of the foot or instep is as nearly as possible the same length.

According to Dr. Sieveking, the weight should be in the following proportion to the height :—

Model Heights and Weights.

The Height being		The Weight should be		
5 feet	1 inch	8 stone	4 lbs.
5 "	2 "	9 "	0 "
5 "	3 "	9 "	7 "
5 "	4 "	9 "	13 "
5 "	5 "	10 "	2 "
5 "	6 "	10 "	5 "
5 "	7 "	10 "	8 "
5 "	8 "	11 "	1 "
5 "	9 "	11 "	8 "
5 "	10 "	12 "	1 "
5 "	11 "	12 "	6 "
6 "	0 "	12 "	10 "

As to this point, see *post*, INSURANCE.

As respects the dimensions of the skeleton at various ages, the following tables by Dr. Humphrey are here given: —

Measurements at different Ages in Inches.

Age.	Height.	Spine.	Circumference of Skull.	Humerus.	Radius.	Hand.	Femur.	Tibia.	Foot.	Pelvis. Transverse Diameter of.	Pelvis. Antero-posterior.
At Birth	19	7.0	15.0	8.5	2.5	3.1	4.3	3.5	3.5	1.3	1.3
2 years (average) .	27	8.5	17.7	4.7	3.6	3.1	6.2	5.1	3.6	2.2	2.2
4 to 6 years (average)	35	11.8	19.0	6.6	4.8	4.1	9.1	7.1	5.1	2.5	2.5
8 to 12 years (average)	43	12.8	18.8	8.3	6.0	5.1	11.4	9.4	6.4	3.1	3.1
15 yrs. { Female .	55	17.0	19.0	10.3	7.0	5.8	14.8	11.0	7.8	4.0	3.6
15 yrs. { Male . .	54	16.5	19.0	10.5	7.5	5.6	15.0	11.5	. . .	. . .	. . .
15 yrs. { Average .	54	16.6	19.0	10.4	7.4	5 7	14.8	11.3	8.0	3.8	3.6
18-19 years. { Female .	59	19.0	19.5	11.0	8.2	6.5	16 0	12 8	8.0	5.0	4.8
18-19 years. { Male . .	59	17.5	20.4	11.0	8.5	6.3	15.0	13.0	8.0	3.9	3.8
18-19 years. { Average .	60	18.5	19.8	11.0	8.4	6.4	15.8	13.3	8.0	4.7	4.5

Average Measurement at different Ages, reduced to a Scale of 100.

Age.	Height.	Spine.	Circumference of Skull.	Humerus.	Radius.	Hand.	Femur.	Tibia.	Foot.	Pelvis. Transverse Diameter of	Pelvis. Antero-posterior.
Birth	100.0	36.84	79.00	18.50	13.20	16 30	22.60	18.50	18.50	6.80	6.80
2 Years . . .	100.0	31.40	65.55	17.40	13.33	11.48	22.94	18.88	13.33	8.14	8.14
4 to 6 years . .	100.0	33.71	51.42	18.85	13.71	11.71	26.00	20.23	14.57	7.14	7 14
8 to 12 years . .	100.0	29.76	43.72	19.30	14.09	11.86	26.51	21.86	14.65	7.21	7.21
15 years . . .	100.0	30.74	35.70	19.25	13.70	10.55	27.40	21.48	14.81	7.03	6.66
18 to 19 years .	100.0	30.83	33.00	19.00	14.33	11.11	26.33	22.16	13 83	7.83	7.50
Adult	100.0	34.15	31.54	19.54	14 15	11.23	27 51	22.15	16.08	8.00	6.61

Orfila's tables differ considerably from those of Dr. Humphrey; they are as follows : —

TABLE I.

(Orfila's First Table.) *Stature Calculated from length of Bones.*

[The Measurements are given in Inches and Fractions of an Inch.]

Length of Bone.		Stature.		
	Inches.	Maximum.	Minimum.	Difference.
Humerus (19 obs.) . .	14.50	68.10	64.50	3.60
Ulna 14 " . .	10.66	70.80	65.66	5.14
Femur 12 " . .	17.75	69.66	64.50	5.16
Tibia 11 " . .	14.21	69.66	64.50	5 16

TABLE II.

(Orfila's Second Table.) *Stature Calculated from length of Bones.*

Length of Bone.		Stature.		
	Inches.	Maximum.	Minimum.	Difference.
Humerus (6 obs.) . .	13 00	73.25	69.75	3.50
Ulna 7 " . .	10.66	73.25	65.00	8.25
Femur 7 " . .	18.10	72.00	67.00	5.00
Tibia 7 " . .	15.00	70.50	65.00	5.50

The differences between these two tables show that too much reliance should not be placed on calculations based upon the relative length of bones. In such calculations it may be added that it is usual to add from 1 inch to $1\frac{1}{2}$ inches for the soft parts.

Race.—Differences of race are chiefly manifested in the skull, which Dr. Tidy divides into three classes : —

1. The prognathous skull of the negro, in which there is a forward prolongation of the jaw, with the *foramen magnum* placed far back.

2. The pyramidal skull of the Esquimaux and of the inhabitants of North and Central Asia, in which there is a peculiar lateral projection of the zygoma, due to the form of the malar bones, rendering the skull lozenge-shaped in appearance.

3. The oval or elliptical skull of the Indo-European or Caucasian, which he regards as essentially symmetrical, having no marked prominences and no undue compressions.

The shape of the skull in some savage or half-civilized races is modified by artificial pressure upon the head during infancy; but it is apprehended that questions relating to such skulls will rarely if ever be submitted to a medical jurist, and definite conclusions as to race from the shape of the skull can rarely be formed.

The capacity of the skull, according to the better opinion, seems to vary in different races; but for details upon this subject as well as measurements of human skeletons in different races, the student must be referred to more voluminous treatises.

Definite conclusions as to race can rarely be drawn from other anatomical peculiarities, if we except the color of the skin of the negro and of other dark races, and the various modifications of the hair. In the negro, however, the feet are wide apart and flat, and the *os calcis* has a remarkable backward projection.

Likeness. — This subject, so far as it relates to questions of legitimacy and paternity, has been already considered.

The determination of questions of identity of living persons by likeness will usually depend upon the application of the ordinary rules of evidence, in which the medical jurist has no special interest. The famous Tichborne case is an interesting case in this connection.

Congenital Peculiarities and Hereditary Diseases. — Congenital peculiarities and hereditary diseases often furnish strong evidence upon the issue of identity. Moles, polydactylism, hypospadias, and epispadias are not unfrequently transmitted through several generations.

Nœvi may often furnish evidence of identity. Hereditary diseases, such as syphilis, etc., may have some bearing upon the question of identity, as may also other congenital peculiarities and diseases, which will readily suggest themselves to the medical jurist.

Cicatrices and Tattoo Marks. — Cicatrices and tattoo marks are often important to be considered in this connection. A scar may often be made more conspicuous by briskly rubbing it with the hand, when the whiteness of the scar, due to its deficient blood supply, will render it conspicuous in comparison with the sound skin. As respects the necessity of a scar following a wound, it may be said that a scar always results from a wound involving loss of substance; scars may not result from slight punctures where the surface of the skin has only been pierced, as by the prick of a lancet or the bite of a leech. And again, where the scar arises from a clean cut with a sharp instrument, especially if in the direction of the muscular fibres, it may be so slight and narrow as to escape notice even on a careful examination. It should be remembered that a scar may affect the epidermis only and not the cutis vera; so that where putrefaction has set in and the skin has commenced to peel, it would be impossible to say that no scar existed during life.

As respects the period of cicatrization of a wound, it may be stated that in the case of simple incised wounds in tissues of average vitality, cicatrization may occur in from 14 to 20 days. The rapidity of the process is, however, dependent upon many circumstances, such as the extent of the wound (the more extensive, the less rapid will be the cicatrization), its nature, whether contused or lacerated (in which case cicatrization will be delayed), its position (wounds of the lower extremities healing more slowly than of the upper extremities), the age and health of the person, and method of treatment.

A recent cicatrix is soft, tender, and pink, or at any rate redder than the surrounding skin ; in the course of one or two or more months, the cicatrix becomes less tender, harder, and of a brownish-white color. As its age increases, it becomes less and less sensitive, hard and thick, white and shining; when a scar has assumed this appearance it is impossible to determine the time when the wound was inflicted, except that it may be safely said that it did not result from a wound inflicted two, three, or even four weeks previously. Where the cicatrix is soft, red, and tender, the probabilities are that it is not of long standing.

As to the inference which may be drawn from the character of the cicatrix it may be stated, generally, that the cicatrix of a straight incised wound, involving no loss of tissue, is usually rectilinear ; the cicatrix of an oblique wound is usually more or less semilunar. Incised wounds involving loss of substance, or contused or lacerated wounds, usually produce cicatrices irregular in outline, with the surface depressed and more or less uneven and puckered; the cicatrix of a stab is usually triangular; the cicatrix of a bullet-wound, where the weapon was fired near the body, is large, deep, and irregular, and the probabilities are that tattoo marks, due to particles of gunpowder being driven under the skin, will be found in the surrounding tissue. Where the weapon was fired from a distance the cicatrix is usually a depressed disc, regular in shape and smaller than the ball that caused it. The cicatrix of the wound of exit, if there is one, will usually be larger and more irregular than the wound of entrance. The cicatrix of a burn will vary somewhat with the form of the heated body applied, while the regularity and contraction of the scar will depend upon the extent and depth of the burn.

Cicatrices produced by caustics have usually regular edges; but this will vary, as in the case of burns, to an unlimited degree; much will depend on the caustic used and the mode of its application. Scars from flogging usually appear as faint white lines extending between little circular pits marking the position of the knots of the lash. Scars resulting from bleeding are white and linear; and where the operation has been properly performed the scar is oblique with reference to the direction of the vein. Marks of cupping usually present themselves as a series of small white symmetrical cicatrices. The cicatrix of an issue is single, rounded, and depressed; that of a seton is double, each mark being linear with a band of lymph connecting the two lines. Blisters, as a rule, leave no scar; occasionally, however, where suppuration results, scars do occur, the extent of which depends on the amount of destruction of the parts to which they are applied. The cicatrix of vaccination is an irregular, flat, slightly depressed, honeycombed scar. Small-pox scars are deep and irregular, occupying a place much below the level of the skin. The cicatrix of a syphilitic abscess, on account of the great loss of substance, is usually very deep. Scrofulous ulcers cause irregular and deeply furrowed scars, with hard, uneven edges. The character of the cicatrices of different ulcers will depend much on their locality and the character of the tissue involved.

The locality of a scar will often afford important evidence as to its character. The cicatrix of a wound in the case of an adult who has ceased growing will be smaller than the wound causing it, and as its age increases will diminish in size and become thicker and whiter; but in the case of a child, on the other hand, a cicatrix will increase in size with the growth of the body; cases are recorded where the length and breadth of a scar have thus become doubled.

As to whether a cicatrix can be obliterated either by lapse of time or by artificial means, it appears to be established that scars resulting from slight incised or punctured wounds where the epidermis only has been injured, or where if the cutis has been penetrated there has been no loss of substance, may in time disappear; this is not, however, either necessarily nor invariably the case. Marks of bleeding may become obliterated after the lapse of two or three years; but on the other hand, there are recorded cases where the scars were distinct after periods of twenty-six, thirty, fifty, or even nearly sixty years. Where, however, a scar results from a skin disease or from a wound involving loss of substance, it is exceedingly doubtful whether the scar can ever be obliterated, although changes may occur resulting in its becoming less distinct.

Tattooing consists in first pricking the skin deeply, and then rubbing into the punctures some coloring substance. The process causes considerable inflammation, lasting commonly two weeks or more. In about six weeks the cuticle scales off, and at the end of about two months the skin assumes its normal character, marked with the device pricked in by the operator. Tattoo marks of irregular outline may be accidentally caused by coloring matters, such as coal-dust, etc., finding their way into a wound. Tattoo marks consisting of scattered dots of bluish color are of constant occurrence in gun-shot wounds where the weapon has been fired very near the person.

As to the durability of tattoo marks it seems established that they may entirely disappear in a small proportion of cases; ten years appears to be the minimum period of such disappearance.

The permanence of tattoo marks depends upon, first, the nature of the coloring-matter; and second, the efficiency of the operation.

1. As to the coloring-matter, — vermilion, indigo, and Prussian blue are the colors most disposed to fade. Cinnabar and common ink are next in permanence; while cobalt and ultra-marine, and especially all carbonaceous material, such as India and China inks, soot, coal-dust, and gunpowder, all of which produce a bluish black tattoo, are by far the most permanent. The color-

ing-matter used in tattooing, even after its natural obliteration, may frequently be distinguished in the contiguous absorbent glands.

According to Dr. Tidy, the color produced by taking nitrate of silver is absolutely indelible.

2. As to the efficiency of the operation,—if the surface of the cutis merely be penetrated the marks are very much more likely to fade than if the punctures are carried into the substance of the skin. Tattoo marks are more likely to disappear from a thin-skinned person than from a thick-skinned one. If the operation is well performed, a tattooed skin may after death be macerated in water for an indefinite time without affecting such marks. The separation of the cuticle by putrefaction does not interfere with an efficient tattoo.

Tattoo marks can be obliterated by artificial means; but such means, resulting in suppuration and the destruction of the skin, will leave scars in the place of the tattoo, as a general rule.

Marks of the Hands and Feet. — Evidence as to footprints and hand-marks is usually given by ordinary witnesses; sometimes, however the opinion of the medical jurist may be required upon this subject. It appears to be settled as respects footprints that the prevailing opinion, that the impression of the foot upon the soil will always correspond with the foot making it, is by no means correct; but on the contrary such footprints may be either larger or smaller than the foot producing them, the exact relationship and size depending upon several causes.

1. The material in which the footprint occurs. Where the impression occurs in sand or in any material com-

posed of minute, freely moving particles, the footprint is usually smaller than the foot; and this, in the case of sand, whether it is dry or moist.

On the other hand, if the impression is made in clay or other material not composed of fine and free particles, the impression is invariably larger than the foot.

2. The impression will depend upon the shape of the boot or shoe worn. If the boot or shoe worn is in any respect peculiar as respects shape, or size, or nails on its lower surface, and the impression corresponds thereto, such evidence will be valuable; but if the correspondence is between the print and a boot not worn at the time the impression was made, it is of little value.

3. The size of the impression will depend upon the rapidity of the progression and the level of the ground. The impression produced by the foot of a person when running is smaller than that of the same person walking; and the impression of the same person standing will be still larger. In a slow walk an impression of the whole foot or the greater part thereof will probably be made, while in running the mark of the heel will be less distinct, and that of the front of the foot more distinct than in walking. In going up-hill there may be only slight evidence of the mark of the heel, while the mark of the ball of the foot may be very distinct. In coming down-hill the impression of the whole foot, but especially of the heel, will generally be well defined.

4. The character, shape, and size of the impression will always depend to a great extent on peculiarities of gait; such peculiarities should in each case be carefully studied.

5. Marks of blood on the floor of a room, if corresponding stains be found on the sole of the boot worn by the prisoner, may furnish important evidence.

Where blood-stains are found on the floor or other portions of the room it is better to remove intact the stained portion of board, plastering, etc.

6. The surroundings of all footprints should be carefully examined. In some cases it may be necessary to take a cast of certain footprints, which may be done by raising the temperature of the impressed ground to about 220° F. by holding over it a warming-pan filled with incandescent charcoal, and then dusting over it powdered stearic acid, which will melt and soak in, taking the exact form of the footprints. This, when allowed to cool, may be detached and used as a mould for the production of a plaster-of-Paris cast.

In a recent number of the "Annales d'Hygiène et Médecine Légale" are given the results of a study by Dr. Masson of the question whether certain marks were made by one and the same foot, and so by one person only. He is of the opinion that it is impossible that two human footprints should closely resemble each other unless made by the same foot. The toes, and especially the great toe, leave marks which should be examined attentively; these, and the outline of the digito-plantar depression, the line which defines the plantar arch, are the data for diagnosis. The conclusions drawn by Dr. Masson are —

1. The dimension and shape of the footprints made by the same foot vary with the attitude taken. 2. The two extreme and characteristic types are represented by impressions made by the foot in walking and standing.

3. The expert called to study the matter of footprints should always take impressions of the foot of the accused, in the act of standing and of walking, and should compare only those which correspond with the same attitude. 4. In connection with the measurements made he should always consider the points which throw light upon the individual characteristics of the foot.

The Teeth. — The condition of the teeth often affords important evidence upon the question of identity. The state of the teeth, and especially irregular dentition, should be accurately noted ; and in important cases a cast of the mouth should be taken by a dentist. Human beings have two periods of dentition. The arrangement of the temporary and permanent teeth is given in Table I. : —

TABLE I.

Arrangement of the Temporary and Permanent Teeth.

		Molars.	Canines.	Incisors.	Canines.	Molars.	
Temporary Teeth {	Upper Jaw	2	1	4	1	2 = 10	} = 20
	Lower Jaw	2	1	4	1	2 = 10	

		Molars.	Pre-Molars or Bicuspids.	Canines.	Incisors.	Canines.	Pre-Molars or Bicuspids.	Molars.	
Permanent Teeth {	Upper Jaw	3	2	1	4	1	2	3 = 16	} = 32
	Lower Jaw	3	2	1	4	1	2	3 = 16	

The periods of the irruption of the teeth are given in Table II. : —

15

TABLE II.

The Periods of the Irruption of the Teeth.

(a) *Temporary Teeth.*

6th or 7th month	two middle incisors.
9th "	two lateral incisors.
12th "	first molars.
18th "	canines.
24th "	two last molars.

(b) *Permanent Teeth.*

6th or 7th year	the four anterior or first molars.
7th "	two middle incisors.
8th "	two lateral incisors.
9th "	first bicuspids or præ-molars.
10th "	second bicuspids or præ-molars.
11th to 12th "	canines.
12th to 14th "	second molars.
17th to 21st "	last molars, or " wisdom teeth."

In determining the age from the state of the teeth, it should, however, be born in mind that there are cases of very early dentition; children have been born with teeth, generally the central incisors; other cases are recorded where dentition was preternaturally late. Dr. Tidy records the case of a woman 70 years of age cutting a canine tooth, and there are cases where adults have never cut their teeth.

The development of the teeth, especially the first set, may be retarded by rickets; syphilis, on the other hand, rather hastens the irruption of the teeth, and especially the first set, and may cause peculiarities in the permanent teeth.

Supernumerary teeth, and third dentition, even, have been recorded.

Hairs and Fibres.—The microscopical examination of hairs and fibres frequently affords important evidence of identity in criminal cases. Thus, the identity of hairs found on mutilated portions of a skull, clutched in the hands of a deceased person, on the clothes of the accused, or imbedded in blood on a weapon, etc., with the hair of the deceased, or the accused, may form an important link in the chain of evidence to convict a person accused of crime.

In cases of bestiality, hairs of the animal will almost always be found adhering to the clothes of the person accused. The examination of the hair about the female genitals, in cases of alleged rape, may reveal the existence of spermatozoa.

Where the question is as to the nature of the hair, it should be first washed in water, thoroughly dried, then soaked in turpentine for some time, and afterwards mounted in Canada balsam; it should then be examined with a medium power of the microscope, the power used depending somewhat upon the nature of the fibre. In every examination of fibres, the comparison should be a direct one between the fibre in question and other fibres, the nature of which is positively known. It will be found convenient to keep a series of hairs of different animals and of different fibres permanently mounted for purposes of comparison. The observer should make himself familiar with the appearance of all ordinary fibres, such as cotton, linen, silk, wool, hemp, etc., and the effect produced upon them by various reagents. It would be an interesting matter to go more fully into the examination of this subject, but want of space forbids; the reader is referred to professed treatises on microscopy.

As respects human hair, its structure and development will be found treated at length in professed treatises upon anatomy, histology, and physiology. In this connection it may be stated that hairs are appendages of the skin, each hair being imbedded in a depression of the skin called a "follicle," fixed at the bottom thereof by dilatation of the hair itself called the "bulb." The portion above the bulb and within the follicle is called the "root," and the portion projecting beyond the surface, the "shaft." The outer or cortical part of the hair is shown, when treated with sulphuric acid, to be composed of spindle-shaped, flattened, angular cells containing pigment granules giving the color of the hair; under the microscope it appears to be tubular, although it is not in fact so. The inner medullary portion consists of granular, nucleated cells, angular or rounded in form, arranged linearly, containing air, which causes them to appear black by transmitted, and white by reflected light. The whole hair is surrounded by a cuticular coat of flat epithelial scales; the varieties of arrangement of this cuticular layer are, in great measure, the cause of the different appearances of the hair in different animals.

The following table, altered by Tidy from Dr. Pfaff's work, gives the diameters of hairs from different parts of the body, and from different animals, in fractions of an English inch: —

	Fractions of Inch.
Down (*lanugo*) from a suckling	$\frac{1}{8888}$ to $\frac{1}{280}$
Down " from a young girl's arm	$\frac{1}{1666}$
Down " from the upper lip of a woman .	$\frac{1}{1428}$
Down of beard (*iulus*)	$\frac{1}{166}$
Hair from a woman's head (*capilli*)	$\frac{1}{434}$
Hair from female pubes	$\frac{1}{166}$

Fractions of Inch.

	Fractions of Inch.
Hair from a man's head (*capilli*)	$\frac{1}{333}$
Hair from axilla	$\frac{1}{166}$
Hair from male pubes	$\frac{1}{333}$
Hair from the eyelashes of a man	$\frac{1}{363}$
Eyebrows	$\frac{1}{200}$
Hair from nostrils (*vibrissæ*)	$\frac{1}{333}$
Hair from moustache (*mystax*)	$\frac{1}{200}$ to $\frac{1}{151}$
Hair from the ears (*tragi*)	$\frac{1}{555}$
Hair from the arm of a man	$\frac{1}{1000}$ to $\frac{1}{555}$
Hair from a man's hand	$\frac{1}{370}$
Pig's bristle.	$\frac{1}{100}$
Hair of fallow deer	$\frac{1}{250}$
Hair of horse	$\frac{1}{310}$
Hair of goat	$\frac{1}{600}$
Hair of fox	$\frac{1}{600}$
Hair of cow	$\frac{1}{600}$
Hair of spaniel dog	$\frac{1}{1100}$
Hair of rabbit	$\frac{1}{1125}$

The characteristic appearance of the hairs of different animals will be found in various works upon microscopy, and also in "Woodman and Tidy's Forensic Medicine."

The question whether a particular specimen of hair is human or not is ordinarily not difficult of answer. In questions as to the nature or identity of hair, the answer should always be based upon comparison with authentic specimens. Microscopic examinations will frequently enable one to say whether the hair has been lately cut, shaved, or violently torn from the body. Infant's hair which has never been cut will be found to taper gradually to a point; after the hair has been cut the ends never regain this taper condition, but remain more or less rounded, and not unfrequently split, terminating in two or more branches. For some days after cutting the hair retains a certain smoothness of section. Hairs pulled out by force generally appear crushed and somewhat frayed; the sheath will, as a rule in such cases, be torn away with the bulb. The condition of the hair, however, lost after fevers and other acute diseases, closely resembles, as regards the condition of the bulb, a sheath thus torn out violently. The condition of the hair may also be modified by certain skin and other diseases.

The color of the hair may be changed in a variety of ways. Light or red hair may be darkened by the use of dyes containing lead, silver, or bismuth; the material in such cases may be detected by chemical analysis. Dark hair may be made light by the application of chlorine water, which, however, is apt to make the hair brittle and rotten. A golden tint may be produced by the use of peroxide of hydrogen.

Where the color has been changed artificially, it will usually be marked by want of uniformity; and if the roots of the hair be examined, the new growth will be seen to be of a different color. The general color of the hair will not correspond with the color of the hair on the pubes or trunk. Certain diseases will, at times, effect a great change in the color of the hair; and there are authenticated instances of its sudden change without an apparent cause, and of its becoming suddenly bleached by grief or fright, although the last-mentioned cases do not appear to be so-well authenticated.

It seems to be settled that both hair and nails may grow for a time after death, owing to the molecular life of the epidermis and the hair-follicles continuing for a time after somatic death.

Limits of Sight and Hearing. — Questions of identity sometimes require the consideration of the limits of sight and hearing. The determination of such questions may be modified by defects of vision, such as hypermetropia, or long sight; myopia, or near sight; presbyopia, aged sight; astigmatism (due to the axes of curvature of the cornea being unequal), color-blindness, and the like. A limit is set to vision, even as respects lofty objects, by the shape of the earth. The following table shows the distance in miles of the farthest visible point that can be seen from the top of a given height, taking into account the effects of refraction : —

Height in Feet.	Distance in Miles.	Height in Feet.	Distance in Miles.	Height in Feet.	Distance in Miles.
5	2.96	150	16.2	3,000	72.0
10	4.18	200	18.7	4,000	83.0
15	5.12	250	20.9	5,000	94.0
20	5.91	300	22.9	6,000	102.0
25	6.61	400	26.4	7,000	110.0
30	7.25	500	29.5	8,000	118.0
40	8.37	700	30.5	9,000	125.0
50	9.35	1,000	41.8	10,000	132.0
60	10.25	1,500	51.0	15,000	162.0
70	11.1	2,000	59.0	20,000	187.0
100	13.2	2,500	66.0		

From this it follows that a man of ordinary stature can be seen on a clear day on level ground at a distance of 3½ miles.

As respects hypermetropia, myopia, etc., the reader is referred for information to professed treatises on the eye.

Color-blindness, or Daltonism, in which red and green or other colors cannot be distinguished, has doubtless been the cause of many railway accidents. According to statistics furnished by Dr. De Fontenay, from the examination of 9,659 persons from 8 years of age upwards, 6,945 being above 16 and 2,714 below that age, 217, or 2.25 per cent, were color-blind. Of 4,492 adult males, 165, or 3.7 per cent, were color-blind; among these, 1,001 belonging to the upper classes showed a per cent of 3.09; while in 3,491 artisans, laborers, etc., the per cent was 3.87. Statistics show that the per cent of color-blindness varies greatly with the employment, age, and sex. Of 6,945 adults above the age of 16, consisting of 4,492 males and 2,453 females, 176, or 2.56 per cent, were color-blind. Among the females, however, there were only 11 cases of color-blindness, or 0.45 per cent. Including all the females examined, amounting to 3,819, there were 16 color-blind persons, or 0.42 per cent; while of 5,840 males, adults and children, there were 201 cases of color-blindness, or 3.44 per cent, all the 16 color-blind females belonging to the working classes. Among 2,714 children from 8 to 16 years of age, 41, or 1.51 per cent, were color-blind, — namely, 1,348 boys with 36 color-blind, or 2.67 per cent; and 1,366 girls with 5 color-blind, or 0.37 per cent. Excluding two cases of violet-blindness, there were 56 cases of red-blindness, 24 of green-blindness, and 135 of incomplete color-blindness. In all the cases both eyes were examined separately and found to be affected.

As to the effects of age on acuteness of vision, Dr. De Guéret finds that —

The acuteness of vision at 50 years has diminished $\frac{1}{8}$
   "   "   60   "   " $\frac{1}{4}$
   "   "   70   "   " $\frac{3}{8}$
   "   "   80   "   " $\frac{1}{2}$

Moonlight, Daylight, Starlight, etc. — The light of the moon, as is well known, varies very much; but it is stated by Dr. Tidy that the best known person cannot be recognized by the clearest moonlight at a greater distance than from 16 to 17 yards. While this may be true as respects recognition of the countenance, it would seem that if there were anything characteristic about the form, dress, gait, etc., recognition might be had at a greater distance.

By starlight only, it is stated by the same author that the best known person cannot be identified further off than from 10 to 13 feet.

Dr. Montgomery states a case where a lady was enabled by the light afforded by a flash of lightning to see distinctly and afterwards identify a man robbing her trunk on a dark night in the cabin of a vessel.

As to the light afforded by the flash of firearms, experiments seem to negative the possibility of recognition by such a light; other experiments by other observers seem to show the possibility of such recognition. Very much would depend upon the nature of the arm, ammunition, the relative position of the parties, quantity of smoke, direction of the wind, etc. Under favorable circumstances, — that is, with a bright flash at a short distance on a dark night, and in the absence of artificial light, and where the wind is such that the smoke does not intervene, — it would seem that in the majority of instances recognition is possible.

The recognition of individuals depends upon various points. Where the person is comparatively near, features, color and arrangement of the hair, prognathism, color of the eyes, etc., are the principal means of identification. Beyond a certain distance, stature, gait, general peculiarities, etc., are the principal means of identification. De Guéret from experiment concludes that the best known persons, even those possessing well-marked personal peculiarities, can be recognized only with difficulty, in broad daylight, at a distance of 100 metres, which is a little more than 109 yards. Beyond 150 metres, or 164 yards, he believes recognition to be impossible. Less known and less remarkable people can be recognized in broad daylight only within a distance of 60 to 100 meters (65 to 109 yards). In the case of people who have no personal peculiarities and are almost strangers, he regards 25 to 30 metres (27 to 33 yards) as about the limit of recognition. Very much would depend, however, as it seems to us, upon the observer; for it is a well-known fact that persons accustomed to using their eyes at long distances can recognize persons at a comparatively great distance.

As to the size of the smallest object which may be seen by the unassisted sight, there is considerable difference of opinion. Carpenter states that the smallest square magnitude, black or white, that can be seen on the ground of the reverse color, is about $\frac{1}{465}$ or $\frac{1}{540}$ of an inch; while particles that powerfully reflect light, such as gold dust, of $\frac{1}{1125}$ of an inch, can be seen by the naked eye by common daylight. Dr. Vincent De Guéret states that objects, to be seen at all, must have a diameter of $\frac{1}{6250}$ of an inch.

Lines may be more easily seen than points; thus, according to Dr. Tidy, opaque threads of $\frac{1}{4900}$ of an inch can be seen by most people by the naked eye when held towards the light. Our own experience is that lines ruled on glass very much smaller than any of the above may be distinctly seen under favorable circumstances by the naked eye. In our judgment it would not be a difficult matter to see a line 1 mikron ($\frac{1}{25400}$ of an inch) in breadth; and Prof. William A. Rogers states that he and Professor Pickering and the assistants at Harvard College Observatory have with the naked eye seen lines ruled on glass which could not have been more than the $\frac{1}{150000}$ of an inch in diameter.

Passing from microscopic objects, Dr. Tidy states that at a distance of one foot a person with normal sight can scarcely see an object less than $\frac{1}{25}$ of an inch; and that at greater distances the size must increase proportionately. Our own experience does not agree with this statement. In order to test the matter, the writer submitted the following-described test to 19 different persons of ages ranging from 17 to 50 and upwards. The test consisted of a piece of black paper approximately one millimetre, or $\frac{1}{25}$ of an inch square, pasted upon a white background; of another similar white square upon a black background, and of a black line approximately one millimetre, or $\frac{1}{25}$ of an inch broad, upon a white background. The cards were hung in a good light, not artificial, and approached from such a distance that they were invisible; and the distances at which they became visible and at which the shape of the squares could first be defined were respectively noted.

The mean distance at which the black square upon the white background became visible was 26 feet, 4 inches; the mean distance at which it could be defined was 5 feet, 10 inches. The mean distance at which the white square upon the black background could be seen was 22 feet, 11 inches; the mean distance at which it could be defined was 5 feet, 7 inches. The mean distance at which the black line upon a white background could be seen was 75 feet. The last test was made by only 14 persons; all the rest were made by 19.

Limits of Hearing. — The velocity of sound in air at 32° F. or zero C., is about 1,090 feet per second. This velocity may be considered as increasing about two feet for each degree of Centigrade. At lower temperatures the velocity is less, and at higher temperatures, greater. In water the velocity of sound is 4 times, and in iron 17 times greater than in air. The intensity of the sound in free air diminishes as the square of the distance from the source of sound. At great elevations the loudness of the sound is considerably diminished.

With the normal sense of hearing, Savart fixes the limits of hearing as between 8 complete vibrations per second, and 24,000, while Helmholtz fixes the limit as between 16 and 38,000, or 11 octaves. This limit varies, however, in different people; and practically, no doubt, will be found to be considerably less than the figures above stated. The distance at which any particular sound may be heard in free air varies with the circumstances and with the observer, and perhaps cannot in the present state of our knowledge be definitely stated. Incredible as it may seem, the writer has usually been able to hear the sound of the impact of a rifle-ball against a target consisting of cloth stretched upon a frame, over which (the cloth) is pasted paper, at a distance of 800 yards.

Limits of the Sense of Smell. — The limits of the
sense of smell will not often come in question in a
medico-legal examination; still, the experiments recorded below, made by Profs. Edward L. Nichols and
E. H. S. Bailey, of the University of Kansas, are of
such interest as to be worthy of presentation to our
readers. The following substances were made use of:
oil of cloves, nitrite of amyl, extract of garlic, bromine,
cyanide of potassium, prussic acid, oil of lemon, and oil
of wintergreen. A series of solutions of each was prepared, such that each member was of half the strength
of the preceding one. These series were extended by
successive dilutions till it was impossible to detect the
substances by smell. The order of the bottles containing these solutions was completely disarranged, and the
test consisted in the attempt to classify them properly
by the unaided sense of smell. The first series of tests
was made by 34 observers, — 17 male and 17 female;
the results of which are indicated in Table I.

TABLE I.

Amount Detected.

	Oil of Cloves.	Nitrite of Amyl.	Extract of Garlic.	Bromine.	Cyanide of Potassium.
Average of 17 males.	1 part in 88,218 of water.	1 in 783,870	1 in 57,027	1 in 49,254	1 in 109,140
Average of 17 females.	1 part in 50,667 of water.	1 in 311,330	1 in 43,900	1 in 16,244	1 in 9,002

In Table II. the same method of investigation was
followed with the following results:—

TABLE II.

Amount Detected.

	Prussic Acid.	Oil of Lemon.	Oil of Wintergreen.
Average of 27 } males.	1 part in 112,000 of water.	1 part in 280,000 of water.	1 part in 600,000 of water.
Average of 21 } females.	1 part in 18,000 of water.	1 part in 116,000 of water.	1 part in 311,000 of water.

Some striking individual peculiarities were developed in the course of these experiments. Three of
the male observers could detect one part of prussic acid
in about 2,000,000 parts of water; two of these persons were engaged in occupations favoring the culti-
vation of this sense. Careful chemical tests failed to
show the presence of prussic acid in several of the more
dilute solutions in which it could be detected by the
sense of smell. It was found that some of both sexes
could not detect prussic acid even in solutions of almost
overpowering strength. There were also several instances of the same peculiarity as respects bromine.
As will be seen from the tables, the averages show
that the sense of smell is in general much more delicate in the case of male than of female observers.

Stains.—The identification of a variety of stains,
such as seminal stains, blood-stains, etc., is often a matter of great importance in criminal trials, and may
conveniently be considered in this place. The subject

of seminal stains has been, perhaps, sufficiently considered in another chapter. The subject of blood-stains will be considered in this place.

In examining stains suspected to be blood, their number, size, shape, location on the apartment, garment, or instrument submitted for examination, should be accurately recorded. Spots of blood will, as a rule, have well-defined and somewhat raised edges; their color will depend on their age and thickness, the moisture and temperature to which the blood has been subjected, and the nature of the material upon which it has fallen. There are three sorts of tests by which its nature may be determined, namely, the microscopic, chemical, and spectroscopic tests.

1. **Microscopic Examination of Blood-stains.** — The examination of a spot alleged to have been made by blood should, in our judgment, be first made with a medium power of from 400 to 500 diameters, for the purpose of ascertaining its general characteristics. All measurements of blood-corpuscles for medico-legal purposes, should be made with as high a power as will give perfect definition. The author is in the habit of using a power of about 1,500 diameters, and is of the opinion that a power of at least 1,000 diameters should be used for such purpose. The technique of such examination cannot be entered upon in detail in this manual. For extended details the reader is referred to Woodman and Tidy's " Legal Medicine," Tidy's " Medical Jurisprudence," and the learned and interesting discussions of the subject by Dr. J. J. Woodard and Dr. J. G. Richardson, which may be found in the London "Monthly Microscopical Journal," vols. 12, 13, and 16.

The general manner of examination may be briefly stated to be, — that a very small portion of the stained fabric or material should be cut off; or if the stain is found on a hard surface, such as glass or steel, a minute fragment should be removed with a sharp instrument, and broken into fine pieces with a sharp knife upon a glass slide, and such material wet with a few drops of $\frac{3}{4}$ of 1 per cent solution of common salt in distilled water, covered with a thin cover-glass and examined with a microscope. For the purpose of making comparative measurements of fresh blood, the method most usually employed is — having first impeded the return circulation of the finger by a thread or rubber band — to prick the finger, remove a small drop to the surface of a slide or cover-glass, and spread the same in a thin film over such surface by the use of a needle or the edge of another slide, allow it to dry, and then immediately examine it with the highest power at the disposal of the observer. If the preparation is intended to be permanent, it may be cemented to the slide in the usual way. If properly prepared it will keep indefinitely.

Corpuscles thus prepared were thought by the late Dr. Joseph G. Richardson to flatten out a little in drying, so as to give an average diameter slightly in excess of that of fresh blood; but the difference in the case of the human blood-corpuscle is, according to his experiments, only that between $\frac{1}{3212}$ and the $\frac{1}{3375}$ of an inch.

The blood-corpuscles of man and of all mammals except the camel tribe are circular, flattened, transparent, non-nucleated cells, presenting concave sides. In the camel tribe the corpuscles are oval, but contain no nuclei. In birds, reptiles, and fish, the corpuscles are also oval, and are distinctly nucleated. The shape of the corpuscles found in examining any suspected stain may therefore afford conclusive evidence as to the general nature of the stain; that is to say, if oval and nucleated, it can be positively affirmed that it is not the blood of a mammal; if circular, it may be as definitely affirmed that it is not the blood of a bird, reptile, or fish. It should be remembered, however, in this connection, that oval corpuscles as well as ordinary circular corpuscles may be rendered globular by treatment with water.

In the examination of blood-corpuscles, when not dried upon the slide as above described, a so-called normal solution, such as the common salt solution already mentioned, or one of the other solutions which may be found described in the larger treatises upon this subject, should always be used.

In the hands of a competent observer there is very little probability that any other bodies will be mistaken for blood-corpuscles; starch-cells, sporules of certain fungi, and the discs found in certain coniferous woods have been considered as the most likely to be so mistaken.

The question as to whether blood under examination can be identified positively as human blood, or as the blood of any particular animal, has been very much discussed. According to the experience of the writer from the examination of 650 corpuscles taken from his own finger, examined at different times, but under as nearly identical conditions as possible, it appears that if the average of a sufficient number of blood-corpuscles

16

is taken, namely, not less than 100, such average will be found to be sensibly constant, or at least will vary within very narrow limits, not much, if any, greater than the sum of the personal and instrumental errors.

Whether, however, another and competent observer would have arrived at the same average from the measurement of the same corpuscles is open to some question. The experience of the author in micrometry leads him to believe that very much of the discrepancy found by different observers to exist between the sizes of blood-corpuscles of the same sort of animals is due to several causes, namely: that the average is taken from the measurement of too small a number of corpuscles; that the eye-piece micrometers were standardized from stage micrometers whose errors were not known, and from too small a number of observations; that too low a power was used in making such examinations, as well as the use of a defective method of measurement; and finally, although perhaps this is not the least source of error, to that personal error which necessarily exists in every refined measurement.

In order to test the relative accuracy of micrometric measurements with different apparatus in the hands of different competent observers, the author recently ruled on a glass slide 15 spaces of approximately .004 and .008 of an inch, and procured the same to be measured by six well-known microscopists, who were instructed to take the mean of at least five measurements of each space and report the same to the author. The result showed, using standard micrometers by the same maker, that the measurements of the same space by different observers varied from zero to .00011 of an inch.

Whether a similar discrepancy will be found to exist in the measurement of spaces of approximately the diameter of a human blood-corpuscle, with a high power, is in process of investigation. If a similar discrepancy shall be found to exist it will go far towards explaining the differences in the size of corpuscles found by different observers. Had a single measurement only been made instead of taking the mean of five, the above discrepancy would, no doubt, have been much greater. This subject seems to demand more attention and more careful examination than it has hitherto received.

According to the present state of our knowledge it appears to be settled that the blood-corpuscles, even in the fresh state, of man, dog, rabbit, guinea-pig, muskrat, monkey, elephant, lion, whale, seal, otter, kangaroo, capybara, wombat, and porpoise, cannot be distinguished from each other by micrometric measurement. With respect to the corpuscles of other animals presenting a greater difference than exists between the corpuscles of the above-mentioned animals and those of man, it seems to the writer, in the light of the investigations above recorded, that it would be extremely perilous to undertake, by mere micrometric measurements alone, to distinguish the blood of man from that of another mammal.

It is possible that further investigation and more extended knowledge of the relation between these different corpuscles, and of the sources of error in micrometric measurements, may enable a careful observer to distinguish human blood-corpuscles from those of some other mammals; but at present it seems somewhat presumptuous.

For full details respecting such micrometric measurements as have been made, the reader is referred, in addition to the works above cited, to an extended series of measurements published in an article by Dr. Moses C. White, in the first volume of the "Reference Handbook of the Medical Sciences," page 587.

It should also be borne in mind in this connection that not only the number but the size of human red blood-corpuscles is changed to an uncertain extent in some diseases, which tends to make more uncertain any conclusions as to the identity of blood derived from the measurement of blood-corpuscles.

2. **Chemical Tests for Blood.** — The consideration of the physiology, pathology, and chemistry of the blood is beyond the scope of this treatise. For details upon these interesting subjects the student is referred to professed treatises upon Physiology and Pathology, and especially to Dr. Charles's "Physiological and Pathological Chemistry."

The color of the blood is due to the presence of a very complex crystalline substance variously called **Hæmoglobin**, hæmato-globulin, hæmatocrystallin, cruorin, and erythrocruorin. This coloring-matter exists in the blood under two forms: the oxidized form, which is of a scarlet color; and the deoxidized form, which is more or less purple. Oxidized hæmoglobin, oxyhæmoglobin, or scarlet cruorin, is found in arterial blood; and the reduced hæmoglobin, or purple cruorin, is found in venous blood combined with more or less oxidized hæmoglobin. The blood found in a dead body, provided access of air to the blood is prevented, derives its color from reduced hæmoglobin and gives the spectrum of reduced hæmoglobin only; exceptions to this post-mortem condition of the blood exist, however, after poisoning by hydrocyanic acid, after death by cold and starvation (in which the reducing powers of the tissue are much diminished), and especially in carbonic-oxide poisoning, where the blood exhibits the particular spectrum of carbonic-oxide hæmoglobin hereinafter described.

Hæmoglobin can be separated from the corpuscles by any means tending to their dissolution,— as, the addition of water to the blood; the passage through it first of a current of oxygen and then of carbonic acid; freezing and subsequent thawing, repeated several times; electrical discharges; agitation of the blood with ether; and the addition to the blood of certain salts or of crystallized bile. By any of these processes the blood is rendered transparent or laky.

Hæmoglobin crystals are obtained with difficulty from human blood, but more readily from that of the guinea-pig, dog, or rat. For the technique of their separation, see Dr. Charles's " Physiological and Pathological Chemistry," page 199.

Hæmoglobin is remarkable for its indiffusibility; it is insoluble in absolute alcohol, ether, chloroform, or benzole; it is readily soluble in water and in weak alcohol, and also in alkaline solutions, but is decomposed both by acids and alkalies, when the body now called hæmatin (also a very complex structure), together with an albuminous principle, is formed. It should be noted, however, that hæmatin is not formed by the action of hydrocyanic acid upon hæmoglobin. Nearly all the iron of the blood is contained in the hæmatin.

Hæmatin is a bluish-black amorphous body, forming a reddish-brown powder, which, when burnt, leaves behind pure oxide of iron. It is insoluble in water, alcohol, ether, and chloroform, but readily soluble in alkalies or alkaline carbonates; it is with difficulty soluble in acetic or the mineral acids, and hydrochloric acid appears to be the only one of these that dissolves it without its iron separating. Solutions of hœmatin are dichroic, being reddish-brown in a thick layer, and olive-green in a thin layer. Like hæmoglobin, hæmatin exists in an oxidized and a reduced condition, freshly reduced hæmatin passing back rapidly into the ordinary form. Hæmoglobin, where the blood has been kept for a long time, becomes changed into hæmatin; and hæmatin, whether produced by age or by chemical actions, like hæmoglobin, in its two states of oxidation has separate spectra.

The time required to change hæmoglobin into methæmoglobin, or hæmatin, varies according to circumstances; it is said to be rapid in towns, but slow in the country, and to be especially rapid where the stain is exposed to an atmosphere in which coal-gas is burnt, any weak acid tending greatly to accelerate it. The change is rapid also where the stained fabric has been worn next to the skin. As a rule, it may be said that if the color of the blood-stain be a bright red, it is proof that the stain is recent; but if it be brown, it is not proof that it is old.

The fact that hæmoglobin is very soluble and hæmatin very insoluble, is of great medico-legal importance. After an article once stained with blood has been washed in water — provided sufficient time has elapsed

for the hæmoglobin to be converted into hæmatin —
enough will in all probability remain to serve for its
identification; but where the stain is perfectly fresh,
and the fabric is washed in cold water before the hæmoglobin has had time to be converted into hæmatin, the
whole of the blood may be so effectually removed by
efficient washing that no trace will remain; hot water,
however, will not effect the removal of a fresh blood-
stain like cold water, owing to its further action on the
coloring-matter of the blood. Where, therefore, in a
criminal case, it appears that an article has been washed
in cold water, evidence of the absence of blood-stains
is of little value; but if it has been washed in hot
water the probability is that the presence of the blood,
if it existed on the fabric, can be satisfactorily demonstrated. The age of the stain is no impediment to the
spectroscopic test hereinafter described. Dr. Sorby has
been able to discover the spectrum of hæmatin after
forty-four years, while Dr. Tidy states that he has
obtained excellent spectra from stains which he had
good reason to believe were over one hundred years
old.

The further consideration of the spectroscopic examination of the blood will be found *post*.

Hæmin (Teichmann's blood-crystals) has not been
found pre-formed in animal bodies; it is a bluish-black
or dark-brown metallic-looking, very staple crystalline
powder. It is almost insoluble in dilute acetic acid,
water, alcohol, ether, and chloroform; but is soluble in
hydrochloric and sulphuric acids and in solutions of
the alkalies and alkaline carbonates, but is decomposed
in its solution. Hæmin crystallizes in numerous forms

belonging to the rhombic system, which are most generally small, brown, or almost black rhombic prisms or
tables.

Teichmann's test for hæmin crystals, as modified by
Buchner and Simon, is substantially as follows : The
stained portion of the fabric or material is to be cut
away from the rest, macerated if recent, and if old,
boiled with an excess of glacial acetic acid till the acid
is colored, when it is to be evaporated to dryness on a
watch-glass or slide. When now examined with a
power of about four hundred diameters, the characteristic crystals of hæmin should be found if the stain is
blood, existing in the form of rhomboidal, tabular, or
needle-shaped crystals, lying across one another in star-
shaped masses, varying in color from a faint yellowish-
red to a deep blood-red. As the presence of the saline
matter of the blood is requisite to the success of this
process, and as this may have been previously all washed
away, a very small particle of common salt may be
added to the acetic acid before the maceration or boiling, in order to insure the appearance of the crystals ;
some advise the omission of the salt as being unnecessary and as liable to encumber the field with crystals
of chloride of sodium ; these, however, can be easily
dissolved out by water, leaving the hæmin crystals un-
touched. To preserve the crystals, they may be sealed
up in acetic acid, or the surplus acid may be removed
and replaced by Farrant's solution.

An alkaline solution of hæmin is dichroic, — brown
by transmitted, and olive-green by reflected light.

Teichmann's test has been shown to be liable to
considerable uncertainty, for the reason that spots of

human blood, or even the fluid itself in appreciable quantity, may fail to yield any hæmin crystals whatever, or only such as are of so indefinite a character as to be utterly worthless for diagnosis. Similar failures have been found to follow attempts at bringing out the polychroism of the blood, even in the hands of competent observers.

In testing a suspected stain, where there is sufficient material, the **action of cold water** upon the stain should be particularly noted; if the stain is recent and upon a material incapable of chemical combination with any of the constituents of the blood, it will be rapidly dissolved by the water, the solution becoming of a rich red or brownish-red color; if the stain is not fresh but still comparatively recent, it is less rapidly dissolved by the water and yields a solution of a dirty-brown color; if the stain is very old, it will be insoluble in water, the hæmoglobin being completely changed into hæmatin.

The chemical tests for blood all have reference to the action of reagents upon its coloring-matter. If the stain be upon a fabric, cut a portion out and treat with cold distilled water; if upon a porous body, such as wood, brick, etc., the stained part should be scraped off for some depth, reduced to a fine powder, and digested for a considerable time in cold distilled water. In either case the liquid should be filtered, and both the matter on the filtered paper and the filtrate preserved for examination. If the stain is upon iron or steel, it may be peeled or scraped off; the scrapings will consist of a mixture of blood and iron. Digest them for several hours in cold distilled water rendered slightly alkaline by ammonia, and should this fail to effect the solution, a trace of citric acid may be used instead. This solution should be filtered; the iron, except a trace of citrate where citric acid has been used, will be left on the paper. The blood solution thus obtained may be tested as follows : —

a. Heat a small quantity in a test-tube to about 149° F; with a blood solution three results will follow: 1. The red color is destroyed. 2. The solution is coagulated. 3. A thick brown precipitate is produced, depending in amount on the strength of the solution.

b. If this brown precipitate is present in quantity, it should be collected upon a filter-paper, dried, and heated with a weak ammonia solution, by which, if blood, it will be soluble. The solution, if sufficiently strong, will appear dark-green by reflected, and red by transmitted light.

c. Where the stain has not been removed from a steel blade, a tincture of galls added to the blood will produce a red precipitate.

d. If the red solution be blood, upon the addition of a very weak solution of ammonia the color will either remain unchanged, or if changed will be slightly intensified or reddened.

e. A solution of chlorine will effect no change on the coloring-matter of the blood, if the chlorine solution be but moderately strong.

f. Strong nitric acid will cause the blood solution to become of a dirty-brown color. If the coagulated mass is sufficient in quantity, heat with strong nitric acid, when a clear yellow solution will be obtained.

g. If a solution of sodic hydrate, ten grains to one

ounce, be added to blood, a dark olive discoloration results, which on treatment with excess of acetic acid changes to red.

Guaiacum Test. — Wet the blood-stain with freshly prepared tincture of guaiacum, and then add a small quantity of an ethereal solution of hydroxyl; if the stain is blood, a characteristic blue tint will be produced. If the material stained is of such a color as to obscure the reaction, add the several reagents and afterwards press the fabric between two pads of white blotting-paper, when the blue color will be absorbed by the paper. In this test the blue color results from the oxidation of the guaiacum resin. It should be remembered, however, that guaiacum is turned blue by a great number of substances, such as gluten, milk, the fresh juice of a variety of roots (such as horse-radish, colchicum, carrot, etc.); also by nitric acid, chlorine, the chlorides of iron and mercury, copper and gold; the alkaline hypochlorites, and a mixture of hydrocyanic acid and sulphate of copper; also by pus, saliva, and mucus mixed with creosote or carbolic acid.

It will be observed that to all the foregoing tests of the existence of blood, except the miscroscopic examination for blood-corpuscles, there are serious if not fatal objections. Moreover, a larger quantity of the coloring matter of the stains is usually required for the satisfactory performance of these tests than is usually found to be present on the stains submitted for examination.

3. **Spectrum Analysis.** — By this test in the hands of a competent observer, the existence of blood in very small quantities may in many instances be determined with great certainty.

The four most important spectra produced by blood are the following: —

1. In the spectrum of **oxyhæmoglobin** the blue end is darkened; two absorption bands are visible in the yellower half of the green, the band nearest the violet end being about twice the breadth of the other band.

2. In the spectrum of **deoxidized hæmoglobin** the blue end is darkened but somewhat less than in the case of oxyhæmoglobin; a single broad absorption band is visible in the green.

3. In the spectrum of blood after short exposure to air (**methæmoglobin or hæmatin**) the blue end is darkened; the two bands of oxidized hæmoglobin are much weakened, and a third band is visible in the red.

4. In the spectrum of **reduced or deoxidized hæmatin**, the blue end is darkened; and two well-defined bands are visible in the green, somewhat nearer the violet than those of hæmoglobin; the band nearest the red end is the narrower, but is intensely black, and has exceedingly well-defined edges. The band nearer the violet is nearly double the width of the other band, but the edges are less distinct; this band may possibly not be seen in very weak solutions.

In the examination of a comparatively recent bloodstain on a white fabric, where there is not a lack of material, cut out a small piece of the stained fabric and soak for a few minutes in a few drops of cold distilled water on a watch-glass; then squeeze out the colored fluid and set the solution by, so that the insoluble matters may be deposited. Fill several experimental glass cells, made of barometer-tubing, with the solution.

1. Examine the aqueous solution with the micro-spectroscope, when, if the blood be tolerably fresh, the spectrum of oxyhæmoglobin above described will be apparent; if such a spectrum is found it is certain that the stain is tolerably recent.

2. Add to the solution in the cell, first, a trace of ammonia, and then a minute fragment of the double tartrate of potash and soda (Rochelle salt); so far no change will appear. Now stir in a small piece of the sulphate of iron and ammonia, with as little as possible exposure of the solution to air, and cover the cell with a cover-glass. The two absorption bands of oxyhæmoglobin will now be replaced by a single intermediate band, fainter but broader than those previously existing, which is the spectrum of reduced hæmoglobin. The hæmoglobin thus reduced may be oxidized by exposure to the air with vigorous stirring, and again deoxidized by further addition of the iron salt. The reduction of the hæmoglobin may be effected spontaneously by merely covering the solution over with a cover-glass and keeping it for some time in the sealed cell. This deoxidation and reoxidation of the hæmoglobin constitutes a very characteristic reaction, and serves to distinguish blood from all other substances.

3. Stir into the solution in a cell a minute fragment of citric acid, which will convert the hæmoglobin into hæmatin, when the bands of the oxyhæmoglobin will disappear; and if the solution be tolerably strong a faint band will appear in the red. If an excess of ammonia be now added the band in the red will disappear, the original bands either not reappearing or, at most, to a very slight extent. If now to the solution in the cell

a very small particle of the double sulphate of iron and ammonia be added, and the solution immediately covered with a cover-glass, in a variable time, say about fifteen minutes, the well-marked spectrum of reduced hæmatin will appear, — the band at the red end being the first to appear. If this solution of reduced hæmatin be exposed to the air and vigorously stirred, the oxidized hæmatin band may often be restored; and also, provided the conversion of the hæmoglobin in the first instance was incomplete, the bands of oxyhæmoglobin.

The late Dr. Joseph G. Richardson, of Philadelphia, who has given great attention to this subject, recommends the following method of examination where the quantity of material is very small. He says :—

"Procure a glass slide with a circular excavation in the middle, called by dealers a 'concave centre,' and moisten it around the edges of the cavity with a small drop of diluted glycerine. Thoroughly clean a thin glass cover, about one eighth of an inch larger than the excavation, lay it on white paper, and upon it place the tiniest visible fragment of a freshly dried blood-clot (this fragment will weigh from $\frac{1}{35000}$ to $\frac{1}{50000}$ of a grain). Then with a cataract needle deposit on the centre of the cover, near your blood-spot, a drop of glycerine about the size of this period (.), and with a dry needle gently push the blood to the brink of your microscopic pond, so that it may be just moistened by the fluid. Finally invert your slide upon the thin glass cover in such a manner that the glycerined edges of the cavity in the former may adhere to the margins of the latter, and turning the slide face upwards, transfer it to the stage of the microscope.

"By this method, it is obvious, we obtain an extremely minute quantity of a strong solution of hæmoglobin, whose point of greatest density (generally in the centre of the clot) is readily found under a $\frac{1}{4}$ inch objective, and tested by the adjustment of the spectroscopic eye-piece. After a little practice it will be found quite possible to modify the bands by the addition of sulphuret-of-sodium solution, as advised by Preyer.

"In order to compare the delicacy of my plan with that of Mr. Sorby, a spot of blood $\frac{1}{10}$ of an inch square may be made on a piece of white muslin, the threads of which average 100 to the inch. When the stain is dry, ravel out one of the colored threads, and cut off and test a fragment as long as the diameter of the filament, which will of course be a particle of stained fabric measuring $\frac{1}{100}$ of the minimum-sized piece directed by Mr. Sorby. When the drop of blood is old a larger amount of material becomes requisite, and you may be obliged to moisten it with *aqua ammoniæ*, or with solution of tartrate of ammonium and protosulphate of iron ; but in the criminal case referred to, five months after the murder I was able from a scrap of stained muslin $\frac{1}{50}$ of an inch square to obtain well-marked absorption bands, easily discriminated from those produced by a solution of alkanet-root with alum and those caused by an infusion of cochineal with the same salt."

In the examination of old blood-stains and of blood-stains on colored fabrics, Dr. Tidy recommends the following procedure : If the blood-stain be old, either citric acid or ammonia (preferably the latter) should be used for dissolving the coloring-matter. If the fabric be colored, that reagent should be employed which possesses the least action on the dye. If the stain, as sometimes happens, be found insoluble in both ammonia and citric acid, it should first of all be acted upon by ammonia, and a moderate heat afterwards applied. The solutions obtained are then to be examined in the manner first above described.

The presence of mordants frequently necessitates some alteration of the procedure, the blood being very likely to become incorporated with the mordant, especially where the fabric has been wet; in such case filtration or allowing the subsidence of the deposit is equivalent to removing the blood coloring-matter. The same details must be carried out as above described and the turbidity of the liquid overcome, not by removing the precipitate but by increasing the intensity of the transmitted light.

In the examination of stained fabrics that have been washed with water after staining, the blood will frequently be found spread over a considerable surface. In such case a large piece of material should be digested with a proportionately large quantity of ammonia or of citric acid, and the solution concentrated by evaporation at a gentle heat, and examined with a micro-spectroscope in the manner already described. In the examination of the water used for washing stained garments or fabrics, it should first be concentrated; if in such concentration any deposit is formed, this should be carefully collected, treated with ammonia, and heat applied if it be insoluble while cold.

Where the stained fabric has been washed with soap and water, the hæmoglobin, by the action of the alkali, will be found converted into hæmatin; in such case there will probably be little difficulty in detecting blood on the fabric by ordinary means. Where it is necessary to examine the soap-water itself, it should be agitated with a large bulk of ether and the mixture

allowed to stand until the ether has well separated.
The ether should then be removed with a pipette and
the residue again shaken up with fresh ether, and this
should be repeated until the aqueous solution is perfectly clear ; the solution remaining is to be concentrated and tested for blood in the usual manner.

Blood-stains on leather, or upon any substance containing tannic acid, require special management on
account of the precipitation of the coloring-matter. In
such case the serum frequently soaks into the leather,
leaving the blood-corpuscles on the surface. Dr. Tidy
directs that in such case a fine slice be shaved off from
the stained portion of the leather, so that there may be
as much blood and as little leather as possible on the
shaving. Bend this shaving so that the stained side
only be brought into contact with water placed in one
of the glass cells already described. In this manner he
says that a solution of the blood coloring-matter may
probably be obtained. Dr. Sorby suggests, however,
that if the leather has been washed after the blood has
dried upon it, it will probably be impossible to obtain
the blood-spectra by this method. The following procedure in such case is said to work satisfactorily : —

Digest the stained leather in a mixture, by measure,
of one part of hydrochloric acid and fifty of water, for
twenty-four hours, which will effect a solution of the
mixed compound of the blood coloring-matter and tannic
acid ; the acid liquid is then to be poured off, but not
filtered. The solution may then appear almost colorless, or of a slight yellow tint. To it add an excess of
ammonia, when the color will become either a pale purple or a neutral tint, the tint-shade being considerably

17

intensified by the addition of the ferrous salt and double tartrate, which are now to be added. The solution is then to be examined in an experimental cell under a light sufficiently intense (such as the lime-light or direct sunlight) to penetrate the turbid solution, when the spectrum of deoxidized hæmatin will become visible. If the liquid be too turbid to allow a direct ray from the sun to be reflected through it, the cell should be placed for a few minutes in a horizontal position so that a little of the deposit may subside; remembering, however, that the removal of the deposit destroys the intensity of the spectrum, — the greater part of the hæmatin existing as a compound insoluble in dilute acid.

In the examination of blood-stains on earth, digest the stained earth for some hours in a considerable quantity of ammonia; pour off and concentrate the solution, and examine the turbid solution with an intense light, such as the lime-light or direct sunlight. A similar process should be adopted in the case of stained fabrics soiled with earthy matters.

In conducting micro-spectroscopic examinations, the following general advice is given by Dr. Tidy : —

1. If the fabric on which the blood-stains occur be colored, the spectrum produced by the coloring-matter, extracted from unstained portions of the fabric, should in the first instance be examined. A little blood may be placed on an unstained portion, and when dry examined with a spectroscope, the object being to determine at the outset the spectrum of the dye itself, and any possible interference likely to result on the blood-spectra.

2. The observer should on no account decide that an observed spectrum from a suspected stain is due to blood

unless it exactly coincides with the bands produced by a known solution of blood of equal strength, treated in the same manner.

3. The spectra should in all cases be examined both by daylight and by artificial light. Direct concentrated sunlight, or the lime-light, should be tried whenever the solution is thick and turbid.

4. Never be content with observing a single spectrum of blood. Remember, further, that it is often impossible to obtain the unaltered blood-spectrum; hence, never be satisfied that a stain is not blood until you have failed to obtain all the spectra produced by the action of appropriate reagents.

5. If the liquid under examination is too strong, so much light will be cut off that the absorption-bands may be obscured; if the solution is too weak the bands will become so faint that they are likely to be overlooked. Solutions of several different strengths should, if possible, be examined.

6. Use excessively minute quantities of the several reagents.

7. Adjust the width of the slit of the spectroscope during the examination. Absorption bands are best defined when the slit is *very* narrow.

8. In the present state of our knowledge, the microspectroscope affords no information whatsoever whether the blood comes from man or beast, or from what class of animals it is derived.

As to whether other substances give spectra similar to those of blood, Sorby says of the spectrum of oxyhæmoglobin: "I do not know of anything that gives exactly the same, but there are some things which give bands so far similar as to show the importance of studying the effect of different reagents."

A form of *chlorophyll*, from the petals of the red variety of cineraria, gives two absorption bands which, though dissimilar in relative width, are nearly similar in position to those of oxyhæmoglobin. With ammonia, however, the absorption bands of blood remain unchanged, while those of cineraria are completely altered. The reds of *cochineal, lac-dye, alkanet, madder*, and *munjeet*, dissolved in each case in alum, while somewhat similar in their absorption bands to blood, are not, when examined side by side with blood, likely to be mistaken for it by the practised observer; all are changed by ammonia, and all are bleached by potassic sulphite, which has no action on blood.

Of all the tests for blood given in this connection, the discovery of red blood-corpuscles by the microscope and the micro-spectroscopic examination above described alone seem to be without fallacy. *In the present state of our knowledge, however, it is impossible to determine whether a given specimen of blood is or is not human.*

Menstrual Blood. — In the case of blood-stains found on the clothes of a female, the question may arise whether or not the blood is menstrual. It has been stated that menstrual blood contains no fibrin, is acid, owing to its admixture with vaginal mucus, and that it is invaribly associated with the pavement epithelium derived from the vaginal walls. Such pavement epithelium, if existing, could readily be demonstrated by microscopical examination; and this last peculiarity might justify an inference as to the source of the blood; but the observer would rarely, if ever, be justified in stating a positive opinion as to the source of the blood from a mere microscopical examination.

CHAPTER XVII.

LIFE INSURANCE.

Expectation of Life, Presumption of Death, etc. — A discussion of the rules of law upon this interesting and important subject would swell the size of this book beyond its prescribed limits. We can, therefore, in this connection, only refer to those topics which are of special interest to the medical examiner, leaving the legal discussion of the subject where it more properly belongs, to professed treatises on the Law of Insurance.

Life Insurance is a contract by which, in consideration of the payment by the insured to the insurers of a certain sum of money called a **premium**, either in quarterly, semi-annual, or yearly instalments or in a gross sum, the insurers agree upon the death of the assured, or upon his arrival at a certain age, or upon his death before that time, to pay either to him, his executors, administrators, or assigns, a certain sum of money. The insurance is sometimes effected for a limited number of years, payment of the sum assured to be made only upon the death of the insured within that period. Insurance is also effected by some companies against partial or total disability by accident.

The writing evidencing this contract is called a **policy**. The policy is based upon an **application** in writing made by the assured, in which full information should be given upon all questions affecting the risk, such as age, occupation, habits, condition of health, disease, family history, etc. The applicant for insurance is also required to submit to examination by a physician, or physicians, as to his physical condition. The answers to the questions propounded by the agent or examiner, as well as the statements made by him in his application for insurance, are warranted by the applicant to be true, and form the basis of his contract with the company.

Life insurance as a business is based upon **the expectation of life.** Various mortality tables have been from time to time computed, — from the first by the astronomer Edmund Halley, from a series of life-registers during the years 1687–91, to the present century. The Actuaries', or Combined Experience Table was published by Actuary Jenkin Jones in 1843, and was based upon the recorded experience of seventeen life companies in England, and was deduced from 62,537 assurances, under the superintendence of a committee of actuaries. It is as follows : —

COMBINED EXPERIENCE TABLE.

Age.	Expectation of Life.	Age.	Expectation of Life.	Age.	Expectation of Life.
10	48.36	40	27.28	70	8.54
11	47.68	41	26.56	71	8.10
12	47.01	42	25.84	72	7.67
13	46.33	43	25.12	73	7.26
14	45.64	44	24.40	74	6.86
15	44.96	45	23.69	75	6.48
16	44.27	46	22.97	76	6.11
17	43.58	47	22.27	77	5.76
18	42.88	48	21.56	78	5.42
19	42.19	49	20.87	79	5.09
20	41.49	50	20.18	80	4.78
21	40.79	51	19.50	81	4.48
22	40.09	52	18.82	82	4.18
23	39.39	53	18.16	83	3.90
24	38.68	54	17.50	84	3.63
25	37.98	55	16.85	85	3.36
26	37.27	56	16.22	86	3.10
27	36.56	57	15.59	87	2.84
28	35.86	58	14.97	88	2.59
29	35.15	59	14.37	89	2.35
30	34.43	60	13.77	90	2.11
31	33.72	61	13.18	91	1.89
32	33.01	62	12.61	92	1.67
33	32.30	63	12.05	93	1.47
34	31.58	64	11.51	94	1.28
35	30.87	65	10.97	95	1.12
36	30.15	66	10.46	96	0.99
37	29.44	67	9.96	97	0.89
38	28.72	68	9.47	98	0.75
39	28.00	69	9.00	99	0.50

The American Experience Table of Mortality is as follows : —

Age.	Expectation of Life.	Age.	Expectation of Life.	Age.	Expectation of Life.
10		38	29.62	57	16.05
20	42.20	39	28.90	58	15.39
21	41.53	40	28.18	59	14.74
22	40.85	41	27.45	60	14.09
23	40.17	42	26.72	61	13.47
24	39.49	43	25.99	62	12.86
25	38 81	44	25.27	63	12.26
26	38.11	45	24.54	64	11.68
27	37.43	46	23.80	65	11.10
28	36.73	47	23.08	66	10.54
29	36.03	48	22.36	67	10.00
30	35.33	49	21.63	68	9.48
31	34.62	50	20.91	69	8.98
32	33.92	51	20.20	70	8.48
33	33.21	52	19.49	71	8.00
34	32.50	53	18.79	72	7.54
35	31.78	54	18.09	73	7.10
36	31.07	55	17.40	74	6.68
37	30.35	56	16.72	75	6.28

The expectation of life, that is, the number of years on an average that a healthy person at a certain age will live, excluding all persons under 25 and over 75 years of age, is represented by the formula of Willich, as follows : —

$$x = \tfrac{2}{3} (80 - a).$$

In which x represents the expectation of life and a the age of the person.

Medical Examination. — The medical examination of an applicant for an insurance should be made by a man of skill and experience, and in by no means as superficial and perfunctory a manner as is usually the case.

The scope of the examination will necessarily vary somewhat according to the requirements of the different companies, — some companies requiring an investigation by the physician of the family history and a personal medical history, as well as an investigation of the applicant's personal condition. Assuming that it is the duty of the examiner to make the more extended examination above described, he should ascertain —

1. **The family history** of the assured, taking care that it is stated clearly and fully, with no ambiguity of terms, and no uncertainty or concealment. General and indefinite statements by the assured regarding deaths should be explained. Symptoms as effects of disease should not be allowed to be stated in the place of the diseases on which they depend. Particular inquiry should be made regarding the following points in the family record: Have there been two cases of apoplexy, paralysis, heart disease or brain affection,[1] or one of each pathologically akin? Have there been two cases of Bright's disease or cancer? Have any two members been insane? Have the questions relative to the final illness of the members of the family who have died been answered particularly as to duration and previous health?

2. **The Applicant's personal medical History.** — The inquiry here should be directed to the point whether his present and past condition warrants the belief that the applicant will reach advanced age, aside from the accidents and contingencies common to all.

[1] For many of the rules here given for medical examiners, the author is indebted to the printed rules prescribed by the Mutual Benefit Life Insurance Company of New Jersey, to its medical examiners.

3. **The Applicant's personal Condition.** — Is he sound in body and mind ? His habits as regards indulgence in spirituous or malt liquors, opium, or tobacco, or in any other stimulant or narcotic, should be carefully investigated, and the organs likely to be injured by such indulgence carefully examined. The temperature should be taken, and if there be heat of skin, it should be observed before he is undressed. The physique and complexion are important facts in determining the character of the risk; a flat chest, pigeon-breast, protuberant abdomen, local muscular atrophy, disproportionate height and weight, excessive height, stooping gait, curved spine, are all serious evidences of impairment. Exact and not approximate measurement should in every case be made. The proper average relation between the height and weight of an individual, according to the tables published by the Mutual Benefit Life Insurance Company of New Jersey and by the Mutual Life Insurance Company of New York (the latter of which was compiled by Dr. Minturn Post and Dr. Isaac Kipp from American lives), is as follows : —

Height.	M. B. of N. J. Weight.	M. L. of N. Y. Weight.
5 feet	120 pounds	pounds.
5 " 1 inch	124 "	120 "
5 " 2 "	128 "	125 "
5 " 3 "	132 "	130 "
5 " 4 "	136 "	135 "
5 " 5 "	140 "	140 "
5 " 6 "	144 "	143 "
5 " 7 "	150 "	145 "
5 " 8 "	156 "	148 "
5 " 9 "	162 "	155 "
5 " 10 "	168 "	160 "
5 " 11 "	174 "	165 "
6 "	180 "	170 "

A variation of 20 pounds at 5 feet and of 40 pounds at 6 feet, and in the same proportion at intermediate heights, will not be considered excessive.

The medical examiner should examine in turn —

1. **The Nervous and Muscular Systems.**

2. **The Respiratory System.** — Too much attention cannot be given to this head. Healthy respiration should be quiet, easy, in the ratio of 1 to 4 or 5 beats of the pulse and not exceeding 20 per minute in adults. The chest should expand freely in all directions,—the muscles of the neck and arm taking no visible part in the act of breathing; the respiratory murmur should be neither harsh nor noisy; drawing a full breath and holding it for a few seconds should cause no distress; an adult should be able to count aloud rather slowly from 20 to 30 without drawing fresh breath; if the blood be well aerated, the lips, ears, and tips of the fingers should present no appearance of a purple or livid hue. In conducting the examination of the respiratory and circulatory systems, the clothes should be removed from the chest, and as much care should be exercised in making the examination as if the physician intended to prescribe for the applicant as a patient.

3. **The Circulatory System.** — The condition of the heart should be particularly examined. The pulse should be regular and not jerking, and neither too compressible nor too hard; its beat should be about 4 or 5 to each respiration, and in the case of the adult, sitting, should not be below 65 or 70, nor above 80 per minute, although there are very exceptional cases of very slow or very rapid pulse with good health. Change of posture should not make any greater difference than about 10

beats per minute. The pulse of the female is slightly more rapid, as a rule, than that of the male sex, — the average number of beats per minute of a healthy female being 75, while that of a male is 70. In infancy the average number of beats per minute is from 120 to 100; childhood, 100 to 90; youth, 90 to 75; middle age, 75 to 65; old age, 70 to 60; decrepit age, 75 to 80.

The beats of the heart should be clear and unattended with any murmur, blowing or rubbing sounds; the first sound should be louder, longer, and lower pitched; the apex beat of the heart should be in the fifth intercostal space, $\frac{3}{4}$ of an inch within and about $1\frac{1}{2}$ inches below the left nipple, and the impulse while plainly perceptible should neither be jerking nor too widely diffused.

Diseases of the heart have practically only two terminations: (1) sudden death, which is the common ending of fatty and brown degeneration, dilatation and atrophy, aortic regurgitation, disease of the coronary arteries, etc.; and (2) dropsy, which is a common ending to most forms of diseases of the heart and its appendages.

4. **The Digestive System.** — Other things being equal those who have good digestion will live the longest, bear the most fatigue, and can best endure the risk of heat and cold, and exposure. Chronic alcoholism makes a decided impression upon this system. In making this examination the state of the tongue and mucous membrane of the mouth, the appetite, regularity of the bowels, presence or absence of symptoms of dyspepsia, size of the liver, color of the skin and conjunctivæ, etc., should be principally noted.

5. **The Genito-urinary System.**—Although an examination of the urine is not required by all companies, it is by some, and by others where the amount insured exceeds a certain sum. In our judgment no medical examination even approximates to completeness without an examination of the urine. The manner of conducting this examination is not within the scope of this work. When made, the examination should be microscopical as well as chemical.

Even when the rules of the company do not require it, certain symptoms may suggest its necessity. The following are bad indications and suggest special inquiry: œdema of the eyelids, backs of the hand, dorsum of the feet, scrotum and vulva; nocturnal micturition; lumbar pains; dysuria; presence in the urine of albumen, sugar, pus, blood, cancer-cells, epithelial and other tube casts from the kidneys, etc. As regards the genital organs, sexual incapacity in males is an early symptom of many neuroses; and in females, the uterus and ovaries are favorite seats respectively of cancer and of cystic disease. In the male, stricture of the urethra, followed as it often is by grave *sequelæ,* is to be regarded as an element of danger.

In making a medical examination for the purposes of insurance much must of necessity be left to the judgment of the examiner; and to undertake to give detailed directions would in most instances be entirely unnecessary; only a few general hints have been here attempted. The examiner, besides the mere personal examination of the applicant, should always bear in mind the influence of habits, place of residence, occupation, climate, etc., on the duration of life.

The influence of pregnancy and child-bearing is important in the case of women. While very few women die while actually in a state of pregnancy, it must not be assumed that pregnancy is a shield for any longer period than that of actual gestation. Again, a considerable number of women die during confinement; of 10,382 women confined for the first time, 168 or one in every 62 died; while of 26,394 multiparæ, 213 or one in every 124 died. There can be no doubt that much of this mortality was due to ignorance and carelessness; but so long as ignorance and carelessness are possible, they must be taken into account in determining the nature of the risk when application is made for insurance.

Presumption of Death. — Where by a policy of insurance the sum insured is payable at death, the burden of proving the death of course rests with the administrators, executors, or those benefited by the death. Death, being proved, is to be regarded as due to natural causes unless the contrary is shown. By the common law, after the lapse of seven years without intelligence concerning the person the presumption of life ceases, and the burden of proof is devolved on the other party. Upon an issue of the life or death of a party, however, the jury may find the fact of death from the lapse of a shorter period than seven years, if other circumstances concur, — as, if the party sailed on a voyage which should have long since been accomplished and the vessel has not been heard from; but the presumption of the common law, independent of the finding of the jury, does not attach to mere lapse of time short of seven years, unless letters of administration have been granted on his estate within that period, which is in such case a conclusive proof of his death. Although the presumption of life ceases at the expiration of seven years from the period at which the person is last heard from, in the absence of proof it will not be presumed that the death occurred at any particular time prior to the lapse of the seven years. In order to warrant such a presumption there must be evidence other than the mere absence of the party without being heard from.

It is said to be the practice of insurance companies in the case of the absence unheard-of of the assured for a considerable time under circumstances leading to the belief of his death, to pay the policy after the lapse of a year or two, in the absence of any suspicious circumstances.

Presumption of Survivorship. — Most treatises upon medical jurisprudence contain more or less discussion upon this subject, nearly all of which has no application whatever to the administration of justice in courts where the common law prevails. Where the succession to an estate is concerned, the question which of two persons is to be presumed the survivor where both perished in the same calamity, but where the circumstances of their death is unknown, is by the common law a matter to be determined by the evidence, no positive rule being laid down upon the subject. These questions are not to be decided by mere presumption, but are to be tried as they arise, like other questions of fact. Courts of probate, equity, and law alike refuse to presume simultaneous death or survivorship in the absence of evidence. This question has been considered in the Roman law and in several other codes; for the full statement of the rules of the Roman law upon this subject, and of those codes which have in this respect followed the Roman law, see 1 " Greenleaf on Evidence," sec. 29 *et seq.*; and 1 "Tidy's Legal Medicine," page 383 *et seq.*

Insanity and Suicide. — The subject of insurable interest and the legal effect of certain conditions to be found in most life policies, the effect of concealment of facts material to the risk, etc., while matters of great interest and importance, do not properly come within the scope of a work on Legal Medicine, but are more properly considered in legal treatises upon Insurance.

Most policies contain provisions relating to the effect of suicide, providing for the most part that the policy shall be void if the insured commit suicide or shall die by his own hand, whether sane or insane. Much litigation has arisen as to the effect of such clauses, but the questions discussed in the recorded cases are questions of law rather than of medicine, and do not come within the scope of this work. If a person is found dead, questions may occur under some policies which will require the aid of the medical jurist in determining whether the death was natural, accidental, suicidal, or homicidal. The burden of proof that the death was not natural would seem to rest with the insurers, and the mode and cause of death, when the question arises, is to be determined upon principles discussed in other chapters. It may be stated in passing, that it is well settled that suicide is not proof of insanity. See *post*, chapter upon INSANITY.

CHAPTER XVIII.

Definitions. — A feigned disease, strictly so called, is one which is altogether fictitious.

A factitious disease is one which is wholly produced by the patient, or at least with his connivance.

A latent disease is a real disease which presents little or no outward manifestation during life, and is only detected on inspection after death.

To this classification some writers add exaggerated diseases, or those which, existing in some degree or form, are pretended by the party to exist in a greater degree or a different form; and aggravated diseases, or those which originate in the first instance without the person's concurrence, and are intentionally increased by artificial means.

1. **Feigned diseases** are simulated from a variety of motives, such as avoidance of military and naval service, the obtaining of a pension, escape from imprisonment or other punishment, obtaining alms from the charitable, obtaining damages in courts of law for simulated iujuries alleged to have been received in railway accidents, and from many other motives too numerous to mention.

18

Simulated insanity will be considered in another connection. Feigned poisoning will be considered under TOXICOLOGY. Pretended delivery has already been referred to.

The diseases best adapted for the purpose of simulation are those of a chronic kind, in which the symptoms are purely subjective and produce no apparent disturbance of the system, and consequently call for less self-denial on the part of the simulator; and owing to their existence resting mainly on the veracity of the patient, are less easily proved to be simulated. Although cases of simulation are more often met with in the army and navy, they are not unfrequently encountered in hospital, dispensary, and private practice.

The consideration of this subject must necessarily consist principally in the narration of particular cases, many of which are to be found in the larger treatises on Medical Jurisprudence. A few only will be stated in this connection.

Dr. Ogston relates the case of a beggar in Aberdeen, who simulated elephantiasis very successfully. The pretended swelling was confined to one leg below the knee, and was produced by padding; to favor the deceit, part of the leg was covered with gold-beater's-skin, and exposed by drawing down his stocking two or three inches.

Epilepsy is a disease frequently simulated by beggars; and next to that, according to Ogston, the loss of an arm. To give color to a feigned attack of epilepsy, a ligature around the neck has been applied to induce reddening of the face; soap in the mouth to imitate froth about the lips, and the gums have been pricked to give the tongue the appearance of having been bitten. It would be impossible, in such cases, for the impostor to imitate the fixing or twitching of the eyeballs, the insensibility of the pupils, the peculiar perturbation of the heart, rigidity of the muscles, and the insensibility of the skin and mucus outlets; while the prodromata and the sequelæ of the attack, if ascertainable, would assist further in determining the nature of the suspected simulation.

Real epilepsy often offers considerable variety in its character, — one attack being a mere momentary loss of consciousness (*le petit mal*), while another may consist of the most violent convulsions (*le grand mal*). Impostors usually simulate the severer type, with the peculiar cry, falling down, struggling, lividity, frothing at the mouth, etc.; but they often omit to feign any of the sequelæ of true epilepsy. Many such cases have been unmasked by threats, or by quietly bringing some sharp or hot substance into contact with some part of the body. Calmeil is said to have detected a simulator, who fell in a pretended fit on a heap of straw in the street, by ordering the straw to be set on fire.

The true epileptic usually falls forward, and thus frequently injures his nose, forehead, chin, or cheeks. As he falls he is deadly pale, not red; tonic convulsions immediately begin, the trunk muscles being nearly always affected as well as the others; the muscular rigidity is not to be overcome, or only with great difficulty: once overcome, the muscles remain flaccid until some time after, or until another fit sets in; the face only reddens after a time, and then the veins of the neck swell, etc. Then clonic convulsions set in; the

pupils are usually dilated and refuse to contract under
the stimulus of a strong light; the sense of smell is also
abolished, and strong aqua ammoniæ may be held to the
nose with impunity. The attack is succeeded in severe
cases by torpor, drowsiness, and confusion of mind,
which last for some hours, and the patient is left very
weak. Impostors usually, as in insanity, overact their
part; their contortions are too violent, and they mix up
all the stages of the complaint, usually disregarding the
sequelæ.

Deafness and dumbness are very frequently simulated,
as well as neuralgia, spinal irritation, muscular debility,
spasms, paralysis, contraction of joints, and rheumatism,
—all of which demand no great self-denial, and but
little cunning on the part of the simulator. In dumbness, impostors seldom attempt more than ceasing to
speak, and in paralysis, ceasing to move about; in both
cases the sudden development and confirmed state of
the alleged disease from the outset should excite suspicion. Dr. Ogston relates the case of a prisoner in the
Aberdeen jail who pretended sudden and total deafness,
and whose imposition was detected by dropping at his
back a huge bunch of keys from a high window, without
his being at all startled or taking any notice of it; while
the same thing, done in a case of congenital deaf-mutism,
caused the prisoner to start in alarm and look around
in all directions except that from which the sound or
impulse had come.

True deaf-dumbness occurs only in congenital cases.
The Abbé Sicard is said to have detected one impostor
by noticing that his spelling of written words was pho-
netic. Feigned deafness may sometimes be detected by

the exercise of a little pious fraud.   Dr. Taylor records
one case where a pauper, feigning deafness, was detected
by the production of a case of surgical instruments
during the conversation of the two surgeons regarding
the immediate performance of an operation upon him.
Some startling statements may be whispered, the countenance, pulse, etc., of the supposed simulator being
watched meanwhile.

Hæmoptysis has frequently been mimicked by first
swallowing the blood of animals and ejecting it in the
presence of witnesses.

Scurvy has been imitated by irritating and pricking
the gums.

The unhealthy hue or yellowness of the skin in
chronic dyspepsia, jaundice, and hepatitis, has been
simulated by the use of emetics and purgatives, by the
use of skin dyes, or by the natural or acquired facility
of ejecting the contents of the stomach at pleasure.

The appearance of amaurosis has been imitated by
the application to the eye of belladonna, hyoscyamus,
or atropine to insure the dilatation and insensibility of
the pupils.   A thorough knowledge of ophthalmic medicine and surgery and some acquaintance with optics are
necessary in order to detect some clever impostors.   An
ophthalmoscopic examination will here prove of great assistance ; the use of atropine for the purpose of paralyz-
ing the accommodation will often be necessary, when it
will be frequently found that the pretender of myopia or
near-sight cannot see the test types with concave lenses
which would suit a real sufferer, and the reverse.
Feigned double vision may usually be detected by colored glasses and the use of prisms.   The simulation of

complete blindness is the most difficult of exposure. Pretended blindness of one eye may often be detected by a prism, placed base upwards or downwards before the sound eye, the double vision produced furnishing the required proof.

Contraction and rigidity of the large and small joints are often affected by soldiers and beggars; and some color is occasionally given to the imposture by keeping the limb at rest by bandages till some stiffness and wasting of the muscles ensue; a feigned dieases may thus occasionally be converted into a factitious one.

Simulated nervous diseases, such as hemiplegia, paraplegia, catalepsy, etc., may frequently be detected by the intelligent use of electricity; in such cases the temperature, change of nutrition, reflex movements, peculiar eruptions, etc., if they exist, may throw light on the solution of the question.

As most impostors have little or no knowledge of many of the diseases which they simulate, the simulation can often be detected by the patient's enumerating incompatible, improbable, or impossible symptoms.

An accurate knowledge of the symptoms, course, and pathology of disease, with watchfulness and attention to the surrounding circumstances, and a few simple tests such as will suggest themselves to the intelligent practitioner, will in most cases disclose the fraud. The resources of medical art have of late years been so increased by the thermometer, microscope, ophthalmoscope, stethoscope, laryngoscope, electrical apparatus, sphygmograph, etc., that the exposure of fraud is now much easier than it once was.

Many amusing instances of the exposure of attempted simulations are to be found in the books. Thus, Paré mentions a beggar who had introduced a long piece of bullock's gut into his rectum in order to imitate *prolapsus ani;* from the bowel which was filled with a mixture of blood and milk, he had learned to press out drops at pleasure. A kick from the foot of the inspector made the gut tumble out and disclosed the imposture.

For further cases the reader is referred to the larger treatises of Taylor, Ogston, and other writers.

2. **Factitious Diseases.** — Scrofulous and other sores, stiffening and contraction of the joints, *fistula in ano, ophthalmia,* cutaneous eruptions, and many other ailments are included in this category. Sores have been produced in the neck simulating scrofulous ulcers, by the use of escharotics, and the deceit strengthened by the application of the juice of euphorbium to favor the swelling and redness of the eyelids, nose, and lips.

Inflation of the scrotum to simulate hydrocele, or of the areolar tissue elsewhere to resemble œdema, would only deceive a novice. Œdema has been produced by ligatures. There are recorded instances of pretenders having so overdone their parts in this respect as to produce gangrene of the arm or leg.

Diarrhœa and dysentery have been simulated by taking strong cathartics.

Ophthalmia may be produced artificially by the application of various irritants to the eye.

Various cutaneous affections have been successfully imitated by the use of a variety of irritants applied to the skin.

Tinea capitis has been imitated by the application of nitric acid to the scalp, previously protected by some fatty substance.

This list might be very much lengthened, but enough has probably been said upon this subject.

In the examination of cases of both suspected ficti- tious and factitious diseases, the inspector should be well acquainted with the disease simulated and with the means which may be resorted to for producing the appearance of morbid actions or discharges. The his- tory of the person will often throw light upon the ques- tion; the suspected impostor should be encouraged to give full accounts of the origin, progress, duration, and symptoms of his disease; to do this correctly will in the majority of cases require more knowledge than is possessed by a non-medical man.

Anæsthetics will sometimes assist in unmasking a fraud; in cases of doubt, however, it is more charitable to assume for the time being that the patient's state- ments are true than to run the risk of mal-treating or neglecting a case of real disease. The possibility of latent disease in a person supposed to be an impostor, should also be borne in mind.

See an interesting case narrated by Dr. Ogston, on page 340 of his excellent work on Medical Jurisprudence.

3. **Latent Diseases.** — It is a fact well known to physi- cians that many disorders, even those of which the presence is commonly indicated by well-marked symp- toms, may in particular cases present, throughout the whole or a greater part of their course, a material deficiency or total absence of their usual external char- acters or symptoms; and on this account they are

frequently confounded with other diseases, or entirely escape observation. As stated by Sir Robert Christison, nothing is more common in the practice of medical jurisprudence than for the expert to find his opinion and conduct embarrassed by sudden death arising in the like circumstances, by the discovery of appearances in the dead body adequate, apparently, to account for death, yet unconnected with any traces of the existence of corresponding disease during life.

In the list of latent diseases, Sir Robert Christison enumerates (with perhaps some latitude in the use of the term latent) apoplexy, cerebral meningitis, cerebral inflammation, pleuritis, pneumonia, pneumothorax, pulmonary tubercle, diseases of the great vessels within the chest, such as aneurism, and affections of the abdomen and spine.

Dr. Ogston remarks that the term "latent disease" applies with most force to such diseases as cerebral meningitis, softening of the cerebral lobes of the brain, and abscess in its substance; instances of which he says he has encountered in practice where no complaint of illness was made until within a few hours, or days at most, of the fatal event.

Some fatal diseases of the heart, aneurisms of the large vessels, and apoplexy most nearly approach what may be called strictly latent diseases.

A full consideration of this interesting subject would occupy too much space; for further particulars the student is referred to Dr. Ogston's work and to standard treatises on the Practice of Medicine.

CHAPTER XIX.

MALPRACTICE.

MALPRACTICE may be considered under two heads: civil, and criminal.

1. **Civil Malpractice.** — The gist of an action for civil malpractice is negligence. The rules of law governing this important subject are well stated by Judge Cooley in his work on Torts. We cannot do better than quote his language : —

"As the promise is not different in the case of the physician and surgeon from what it is in the case of the attorney, solicitor, and proctor, one general rule may be given which will apply to all.

"The English authorities are, perhaps, somewhat more indulgent to the faults and mistakes of professional men than are those of this country. Thus, Lord Campbell, with the full concurrence of his associates in the House of Lords, declared that in order to maintain an action against one's legal adviser, it was necessary, 'most undoubtedly, that the professional adviser should be guilty of some misconduct, some fraudulent proceeding, or should be chargeable with gross negligence or with gross ignorance. It is only upon one or the other of these grounds that the client can maintain an action against the professional adviser.'

"On the other hand the rule is laid down in Pennsylvania that the professional man must bring to the practice of his profession a degree of skill and diligence such as those 'thoroughly educated in his profession ordinarily employ.' This is a severe rule, and fixes a standard of professional skill and attainments which, in the newer portions of the country, would be quite out of the question. In New Hampshire the undertaking of the practitioner has been stated in the following language : ' By our law a person who offers his services to the community generally, or to any individual, for employment in any professional capacity as a person of skill, contracts with his employer: 1. That he possesses that reasonable degree of learning, skill, and experience which is ordinarily possessed by the professors of the same art or science, and which is ordinarily regarded by the community and by those conversant with that employment as necessary and sufficient to qualify him to engage in such business. 2. That he will use reasonable and ordinary care and diligence in the exertion of his skill and the application of his knowledge to accomplish the purpose for which he is employed. He does not undertake for extraordinary care or extraordinary diligence, any more than he does for uncommon skill. 3. In stipulating to exert his skill and apply his diligence and care, the medical and other professional men contract to use their best judgment.' This is believed to be an accurate statement of the implied promise. The practitioner must possess at least the average degree of learning and skill in his profession in that part of the country in which his services are offered to the public ; and if he exercises that learning and skill with reasonable care and fidelity, he discharges his legal duty."

A physician possessing a reasonable degree of learning and skill, and exercising according to his best judgment reasonable and ordinary care and diligence, is not liable for a mere error of judgment in advising a particular

remedy about which there is a difference of opinion.
To hold a physician liable for every error of judgment
made in the ordinary course of practice would be to
debar him from the exercise of his profession, and to
deprive the public of the benefit of valuable service.
The law does not require infallibility. A medical man
cannot, as a rule, be held guilty of negligence for not
employing any particular remedy, since, as a rule, there
is never any one specific remedy in the use of which all
authorities are agreed. If it could be shown, however,
that all authorities agreed that a particular drug should
be used in a particular case, as, for example, a certain
antidote in a case of poisoning, the failure to employ
such specific would probably constitute actionable
negligence.

A physician may decline a case, but having once
undertaken it, he must continue his services, even if
gratuitous, until a reasonable time has been given the
patient to procure another physician ; where his services
are not gratuitous, he has no right to desert a patient
without reasonable cause before the end of the illness
he has undertaken to treat.

As respects **voluntary services**, Judge Cooley lays
down the rule thus : "Where friends and acquaintances
are accustomed to give, and do give, to each other voluntary services without expectation of reward, either
because other assistance cannot be procured, or because
the means of parties needing help will not enable them
to engage such as may be within reach, the law will not
imply an undertaking for skill, even when the services
are such as professional men alone are usually expected
to render. And where there is no undertaking for skill,

the want of it can create no liability. So the 'street opinion' of an attorney, given in answer to a casual inquiry by one to whom he holds no professional relation, cannot, however erroneous, render him liable. But when one holds himself out to the public as having professional skill, and offers his services to those who accept them on that supposition, he is responsible for want of the skill he pretends to, even when his services are rendered gratuitously." Under such circumstances, the one who undertakes the treatment of a patient, either voluntarily or upon request, is only liable for gross negligence; but if by forcing himself into a case he excludes a competent physician, he is liable for slight negligence, or for lack of the skill and diligence of the specialist. In general it may be said that the liability of one rendering medical services is measured by the amount of skill he undertakes to exercise; and, as we have seen, the matter of compensation is immaterial.

A physician must always use his best judgment; and while he is not responsible for mere **errors of judgment** or mere mistakes in matters of reasonable doubt and uncertainty, if the error of judgment is so gross as to be inconsistent with the use of the degree of skill required by the law he will be liable to an action. Where errors of judgment result from the want of ordinary care and skill, responsibility attaches, however carefully the judgment is exercised. In exercising his best judgment the physician is only required to anticipate the nature and probable consequences of his treatment. It has been held that he cannot be held responsible for the disastrous effect resulting from

administering chloroform as an anæsthetic to a patient
of a peculiar temperament where such peculiarity was
unknown to him; but in view of the many fatal cases
resulting from the use of chloroform, and the fact of
the uncertainty of its action in this respect, and the
further fact that there are other efficient anæsthetics,
which are safe as compared with chloroform, we cannot
see how at the present time a physician or surgeon can
justify himself in the use of chloroform under ordinary
circumstances. We do not now refer to the careful administration of chloroform in obstetric cases, nor where
the patients are children; there may be some other exceptional cases where its use will be justified. Its administration without a previous careful physical examination
of the patient, or its administration by a dentist without the assistance of a competent physician, and to a
patient in an upright position, seems clearly culpable.
Common prudence would also require that neither it nor
any other anæsthetic should be administered without
the aid of a competent assistant.

A physician cannot lawfully try experiments upon
his patients to their injury.

One who professes to adhere to a particular **school
of practice** must come up to its average standard, and
must be judged by its tests and in the light of the
present day.

As to the mode of treatment in a given case, when it
conforms to the settled practice of the particular school
to which the physician belongs he is relieved from all
responsibility; in such case, evidence of the practice of
physicians of other schools is inadmissible. Where,
however, the case will admit of but one mode of treat-

ment, the use of a different mode would be evidence tending to show a want of skill. The proper and only mode, however, of showing want of skill on the part of the defendant is by proving that he did not exercise it in the particular case. It can neither be established nor disproved by showing the defendant's general professional reputation; it is improper, therefore, to ask a witness what the reputation of the physician is in the community and among the profession as being an ordinarily learned and skilful physician. The treatment of each individual case is the criterion for ascertaining the physician's liability. It has been held, also, that the general opinion of the practitioner with whom the physician studied his profession, or of the professors of the school at which he graduated, are inadmissible.

A sign, or other proof that one actually practises medicine or surgery, is *prima facie* evidence of his professional character. The possession of a medical diploma, issued by a college having authority to grant degrees in medicine, is *prima facie* evidence of ordinary skill.

The mere fact that the defendant refused consultation with other men of his science is no assumption upon his part that he is possessed of more than ordinary skill; and his declination in this respect does not vary the application of any of the rules above stated.

As to the proof of the alleged malpractice, it has been held that the limb upon which the alleged malpractice occurred cannot be exhibited to the jury after the lapse of several years.

It is the duty of the patient to co-operate with the physician in his endeavors to effect a cure. If the injury complained of is due to the contributory negligence of the plaintiff, no recovery can be had; if the want of co-operation of the plaintiff merely aggravates the effect of malpractice on the part of the physician, it will not debar a recovery, but will mitigate the damages. Negligence of the nurse concurring with that of the physician has been held to be imputable to the patient. This, however, would depend upon the question whether such nurse was the agent of the patient or physician. If the acts of negligence can be so separated as to show that the injury was due solely to the fault of the physician, and the fault of the nurse was only remotely connected therewith, an action will lie against the physician.

A physician is responsible for the results of his negligence or unskilfulness notwithstanding the case is given over by him to another; of course, however, he will not be responsible for the subsequent negligence of the physician to whom the case is transferred.

If a family doctor, or the surgeon of a company or society, on leaving home recommends in case of need some other physician, who is not, however, in any sense in his employment, it does not make him in any way liable for injuries arising from the latter's want of skill.

In an action brought by a father against two medical men to recover damages for advising him to suck a tracheotomy-tube of his child, who had just been operated upon for diphtheria, without first duly warning him of the risk he ran, whereby he himself became infected with the disease, it was held that there was no cause of action.

The capacity of the patient injured to judge of the probable results is an important element in cases of malpractice; hence, if the patient is insane he cannot be chargeable with contributory negligence; but where a patient relies upon his own judgment, and not that of the surgeon, as to the propriety of the operation, it is held that the surgeon is not liable for injurious consequences resulting therefrom, — premising, of course, that due care and skill are exercised in performing such operation.

If a person is attacked by a fatal disease and there is no escape from it save by a dangerous surgical operation, then if he gives his free and intelligent consent to the operation and it is skilfully performed, the surgeon cannot be blamed even though the patient perishes under the knife. If a woman in labor is in such a condition that her life can be saved only by the sacrifice of that of the child, then it is not only the right but the duty of the attendant to save the mother at the expense of the child. Of course in such a case counsel should be had when possible before resorting to such extreme measures.

The burden of showing that the use of instruments to produce abortion was necessary to save the life of the woman is on the defendant.

The rules of law above stated respecting the liability of physicians for negligence and malpractice are equally applicable to similar charges made against midwives, nurses, medical students, chemists, pharmacists, and against any other person who holds himself out as possessing special knowledge or skill in any particular department of learning or practice.

Dentists likewise are subject to the same rules as to negligence as physicians and surgeons. A patient must

19

exercise ordinary care and prudence; so that if one tells
a dentist to pull out a tooth, but does not say which
one should be pulled, and the wrong one is taken out,
the sufferer has no legal ground of complaint unless,
indeed, it is quite apparent which is the offending member. The patient may have been a little careless and
negligent, still if the dentist has been so very neglect-
ful of his duty that no ordinary care on the part of the
patient would have prevented the mistake or injury
complained of, the injured party may recover damages.
The fact that one has taken chloroform will not affect
his rights or remedies against a dentist for any mistake
or negligence.

The fact that a dentist extracts teeth gratuitously
does not relieve him from liability for failure to perform
his work properly.

As in the case of physicians so in the case of the
dentist, it is a good answer to an action to recover
payment for his work and labor that the defendant
has been injured instead of benefited by the plaintiff's treatment, either because of his negligence or
want of skill.

Sex is no excuse for negligence, and there is no rule
of law that less care is required of a woman than of
a man.

If a physician should be so indiscreet as to make a
special contract to cure a person of a certain disease or
deformity or to bring about any other desirable result,
he will of course be liable for damages for a breach of
contract if he fails to perform his agreement. Common
prudence would therefore dictate great caution in this
respect, and a guarded prognosis in every case, lest the

patient may pervert what is intended as a mere prognosis into a positive engagement to cure.

Although where the statute of a State requires a State license in order to authorize a physician to practise medicine or surgery an unlicensed physician cannot recover at law compensation for services rendered by him, the fact that he is unlicensed does not affect his liability to an action for malpractice.

Although it may not afford much consolation to the defendant in a malpractice suit, the rule of law is that an action for malpractice does not survive the death of the defendant, and hence does not constitute a claim against his estate.

Although it is not within the power of a State legislature to discriminate in favor of any particular school of medicine, yet such laws may be enacted as will protect the people from ignorant pretenders, and require learning and skill in the school of medicine which the physician professes to practise. No school of medicine is exempt from liability to an action for malpractice.

If a patient voluntarily employs in one art a man who openly exercises another, his folly has no claim to indulgence. The old Mahomedan case cited by Puffendorf with approbation is very much to the point: A man who had a disorder in his eyes called on a farrier for a remedy, who gave him one commonly used upon his quadrupedal patients. The man lost his sight and brought an action against the farrier for damages, but the judge held that no action would lie, for if the complainant had not himself been an ass he would never have employed a horse-doctor.

But on the contrary it has been held that an expert
in the diseases of man is necessarily an expert in the
diseases of animals, so as to make his opinion competent evidence upon the question as to whether a
disease with which a mule was afflicted was of recent
origin or of long standing.

However much a regular physician may differ from the
method of treatment adopted by the profession of any
other school, such as homœopathy or hydropathy, these
systems have acquired a certain recognition by the public, and in some instances legal recognition by statute ;
and it is not to be expected that an adverse verdict will
be given merely because the medical attendant *pro
tempore* practised according to these systems. It is a
well-known fact that the same drugs are to a very considerable extent used by the best practitioners in the
two leading schools of practice ; and so far as concerns
the practice of surgery and obstetrics, the general principles of practice must necessarily be the same in every
rational system.   On the other hand, — as is well observed by Drs. Woodman and Tidy in their manual of
" Forensic Medicine," — mesmerism, Coffinism, Morri-
sonism, have never acquired such a status, and are not
likely to do so.

Certain other systems of charlatanism which at present are occupying an undue share of the attention of
the more credulous portion of the community, such as
the so-called " faith healing " and " metaphysics," might
well be included in the same category. It would, as it
seems to us, be very difficult to frame a defence to a
charge of malpractice preferred against one professing
to heal certain ailments by either of the last two

so-called systems. Suppose the case, for instance, of the treatment of strangulated hernia, or a compound fracture, dislocation, *placenta prævia*, or hemorrhage from a divided artery, by means of faith or so-called metaphysics, or any other like inefficient means. It would not be possible, in our judgment, to frame a defence which would relieve a party practising such or any other mere expectant treatment from liability for malpractice.

As to the question of **doses**, however, even in so-called orthodox medicine, very considerable latitude within certain extremes must necessarily be allowed. Common-sense and experience are the only safe criteria in such case. As is well observed, however, by Drs. Woodman and Tidy, "No experience and no theory can be held to justify giving an infant of a month old $\frac{1}{2}$ grain doses of opium at frequent intervals as a dental sedative, or $\frac{1}{4}$ grain doses of strychnia as a tonic."

Several rules may be found in treatises upon therapeutics by which to determine the dose proper for patients under adult age.

Dr. Young's rule is to add 12 to the age of the patient and divide the age by the sum. Thus, a child one year old would require $\frac{1}{13}$, and one 3 years old $\frac{3}{15} = \frac{1}{5}$ of the amount necessary for an adult.

By Dr. Cowling's rule the proportionate dose for any age under adult life is represented by the number of the following birthday divided by 24. Thus, for a child one year old the dose would be $\frac{2}{24} = \frac{1}{12}$ of that for an adult.

Professor Clarke's rule is based upon relative weights. Assuming the average weight of an adult to be 150 lbs., for whom an appropriate dose is unity, the dose of most medicines must be increased or diminished in the proportion of the weight of the patient to that number of pounds. This proportion is represented by a fraction whose numerator is the patient's weight and whose denominator is 150.

Actions for malpractice are most frequently brought where an unfavorable result has followed the reduction of a fracture or dislocation, or some other surgical operation. It would be interesting to review the reported cases wherein these particular acts of malpractice are more fully discussed, and also to state the leading principles of the practice of medicine and surgery bearing upon such cases; but the limits assigned to this work forbid entering into such details. For further details the student is referred to "Woodman and Tidy's Forensic Medicine," and to the "Compilation of Reported Cases upon Civil Malpractice" by Dr. McClelland, in which latter book will be found a large number of reported cases of malpractice in the various departments of medicine, surgery, obstetrics, etc.

2. **Criminal Malpractice.** — Mr. Bishop, in his work on Criminal Law, vol. i., § 314 (7th ed.), lays down the rule as to homicide from carelessness thus: —

"Every act of gross carelessness, even in the performance of what is lawful, and *a fortiori* of what is not lawful, and every negligent omission of legal duty, whereby death ensues, is indictable either as murder or manslaughter.

"If a man take upon himself an office or duty requiring skill or care, — if, by his ignorance, carelessness, or negligence, he cause the death of another he will be guilty of manslaughter. . . . If a person, whether a medical man or not, profess to deal with the life or health of another, he is bound to use competent skill and sufficient attention; and if he cause the death of another through a gross want of either, he will be guilty of manslaughter."

In vol. ii. of his work on Criminal Law, § 664, Mr. Bishop says: —

" The doctrine as to physician and patient is not quite the same in England and the United States. And possibly it is not entirely harmonious among our States. According to English adjudication, whenever one undertakes to cure another of disease, or to perform on him a surgical operation, he renders himself thereby liable to the criminal law if he does not carry to this duty some degree of skill, though what degree may not be clear; consequently, if the patient dies through his ill-treatment, he is indictable for manslaughter.

" Still, in an English case [says Mr. Bishop in the same section], Willes, J., once put the doctrine in a more reasonable way, thus: 'If a man *knew that he was using medicines beyond his knowledge,* and was meddling with things beyond his reach, that was culpable rashness. Negligence might consist in using medicines in the use of which care was required, and of the properties of which the person using them was ignorant. A person who so took a leap in the dark in the administration of medicine was guilty of gross negligence.'" Mr. Bishop then very characteristically and somewhat dogmatically observes: " Now, in the facts of human life, the less a man understands of anything occult, like the unseen workings of medicine, the more confident he

is that his knowledge of the thing is perfect. Therefore some of our American courts have laid down the
doctrine, not altogether inharmoniously with this utter-
ance of the learned English judge, in substance, that,
since it is lawful and commendable for one to cure another, if he undertakes this office in good faith, and
adopts the treatment he deems best, he is not liable to
be adjudged a felon, though the treatment should be
erroneous, and in the eyes of those who assume to know
all about this subject, which in truth is understood by
no mortal, grossly wrong; and though he is a person
called by those who deem themselves wise, grossly ignorant of medicine and surgery." 2 Bish. Cr. Law, § 664,
citing *Commonwealth* vs. *Thompson*, 6 Mass. 134; and
Rice vs. *State*, 8 Mo. 561.

We have quoted thus largely from Mr. Bishop's excellent book because we believe that in thus dogmatiz-
ing concerning the lack of knowledge of their profession
by those following the practice of medicine and surgery, he has himself erred through insufficient knowledge. "If a man *knew that he was using medicines
beyond his knowledge*, and was meddling with things
beyond his reach, that was [indeed] culpable rashness." But it does not appear that Mr. Justice Willes,
in the case from which the above quotation was made,
which was from a charge to the jury at the Durham
Assizes, 1864, intended to say that this was the only
kind of culpable rashness. It seems, on the other hand,
that this was merely an illustration; for he immediately
adds : "Negligence might consist in using medicines
in the use of which care was required," etc. See *supra*.
That it was merely an illustration is further apparent

from the fact that immediately thereafter he adds
another illustration: "If a man were wounded, and
another applied to his wound sulphuric acid, or something which was of a dangerous nature and ought not
to be applied, and which led to fatal results, then the
person who applied this remedy would be answerable,
and not the person who inflicted the wound, because a
new cause had intervened." In the beginning of his
charge the learned judge very properly said: "Every
person who dealt with the health of others dealt with
their lives, and every person who so dealt was bound
to use reasonable care and not to be grossly negligent.
. . . Another sort of gross negligence consisted in rashness, where a person was not sufficiently skilled in
dealing with dangerous medicines which should be carefully used, of the properties of which he was ignorant,
or how to administer a proper dose. A person who
with ignorant rashness and without skill in his profession used such a dangerous medicine acted with gross
negligence." The drug given in this case, and which
caused death, was a tablespoonful of a tincture of col-
chicum-seeds, containing eighty grains of the seeds, —
eighteen grains, as is said in this case, being a fatal
dose.

In *Nanny Simpson's Case,* 1 Lewin, 172, 262, the prisoner was indicted for manslaughter in having caused
the death of a man by administering *white vitriol* as a
medicine. Bailey, J.: "I am clear that if a person not
having a medical education, and in a place where persons of a medical education might be obtained, takes on
himself to administer medicine which may have a dangerous effect, and such medicine destroys the life of the

person to whom it is administered, it is manslaughter. The party may not mean to cause death; on the contrary, he may mean to produce beneficial effects; but he has no right to hazard medicine of a dangerous tendency where medical assistance can be obtained. If he does, he does it at his peril." See also *Tassymond's Case*, 1 Lewin, 169, where the prisoner was convicted of manslaughter in causing the death of an infant by negligently selling laudanum for paregoric.

It may be conceded that the cases of *Commonwealth vs. Thompson*, 6 Mass. 134 (decided in 1809), and *Rice vs. State*, 8 Mo. 561 (decided in 1844), — in the former of which the law of the case is contained in Chief Justice Parsons's charge to the jury that tried the prisoner, and the latter of which is apparently decided mainly upon the authority of the former, — seem to lay down the rule that in order to warrant a conviction for murder or manslaughter, the defendant must have some knowledge of the fatal tendency of the prescription. An attentive perusal of these cases cannot fail, as it seems to us, to convince the reader that there was a palpable failure of justice in both cases.

In the case of *Commonwealth* vs. *Thompson* the defendant gave to the patient suffering with a cold, powdered lobelia, and persisted in giving it to him for a period of eight days till he was so completely exhausted that no relief could be afforded, and he died of exhaustion.

In *Rice* vs. *The State* the defendant in the court below was employed by the husband of a woman near the end of the eighth month of pregnancy, to cure her of "sciatica;" and after having been informed of her condition and that other physicians had cautioned against the use of vapor baths and emetics in her then condition, he commenced a course of treatment by steaming and giving lobelia, and persisted in this treatment till she had a premature delivery, a few days after which she died. The evidence showed that she had been married five years, and during that time had had three children, always doing well after a confinement, and was in better health when the defendant commenced his practice on her than she had been for many years.

After a consideration of the reported cases it seems to the writer that the cases of *Commonwealth* vs. *Thompson* and *Rice* vs. *The State* are clearly wrong. In our opinion the rule laid down in the late case of *Commonwealth* vs. *Pierce*, decided by the Supreme Court of Massachusetts, is much more rational, and virtually overrules the rule laid down in *Commonwealth* vs. *Thompson*. In *Commonwealth* vs. *Pierce* it was held — Holmes, J., delivered the unanimous opinion of the Court — that to constitute manslaughter where there is no evil intent, it is not necessary that the killing should be the result of an unlawful act; it is sufficient if it is the result of reckless or foolhardy presumption judged by the standard of what would be reckless in a man of ordinary prudence under the same circumstances.

The defendant in this case, who publicly practised as a physician, being called upon to attend a sick woman, caused her with her consent to be kept in flannels saturated with kerosene for three days, by reason of which she died. There was evidence that he had made similar applications with favorable results in other cases, but that in one the effect had been to blister and burn the flesh, as in the present case.   It was held that the jury having found that the application was made as the result of foolhardy presumption or gross negligence, a conviction of manslaughter was proper.

The court in the case of *Commonwealth* vs. *Pierce* has carefully limited the application of the rule there laid down to cases where there was no sudden emergency and where no exceptional circumstances were shown; and thus limited, the rule of the case seems eminently reasonable and grounded on the soundest views of public policy.

CHAPTER XX.

Toxicology treats of the history and properties of poisons and their effect upon the living body. Although this subject is commonly treated as a part of medical jurisprudence, it has practically become a distinct science; and at least so far as concerns the chemical analysis of the stomach or other viscus for the detection of poison, he would indeed be a bold if not reckless man who would undertake to conduct such analysis for medico-legal purposes without special technical training greater than is possessed by the average medical practitioner.

· The growth of the science and the number and importance of the facts connected therewith are so great, as to demand more time and attention in order to render one expert than can be devoted to the subject by a medical practitioner. We have thought it best, therefore, to include in this volume only the subject of general toxicology, by which we mean such consideration of the subject as is necessary to enable a medical practitioner, who usually is the first to have knowledge of the facts in any alleged case of poisoning, to so conduct the preliminary examination that the subsequent and more particular chemical examination by an expert chemist may not be vitiated by the negligence or ignorance of those who have preceded him. It would be better, perhaps, if both the preliminary inspection and the subsequent chemical analysis could be conducted by one and the same person; but considering the magnitude of the subject, and the tendency to specialization which exists at the present time, this is perhaps not to be expected.

From the definition of toxicology above given it is evident that a correct understanding of the subject will involve an intimate acquaintance not only with the science of chemistry, but also with materia medica, semiology, physiology, therapeutics, and pathology. It will also at times include within its scope the examination and investigation of fictitious or adulterated articles, medicines, food, drink, pigments on walls, dyes on clothing, etc.

Notwithstanding the immense accumulation of learning upon the subjects above mentioned, it must be confessed that, in not a few instances, it is very difficult if not impossible to discriminate with certainty during life between poisoning and ordinary disease; and in an equal number of cases it will be found that the inspector is also incapable of laying down any characteristic distinctions between the appearances left in a dead body by certain diseases and those which result from the action of poison on the tissues. Moreover, when we come to the evidence derived from chemical analysis, it will be found in the case of the organic poisons that, in the present state of our knowledge (or rather ignorance), it is impossible thus to determine the existence in the body of some such poisons.

The subject of **ptomaines**, or certain basic bodies developed during putrefactive decomposition of animal substances, moreover, opens up a new field of investigation, and one which has thus far been comparatively little cultivated. Certain poisonous principles have also been found in certain pathological conditions of the body, and even according to some observers, in normal tissues and secretions, the exact conditions of the development of which do not appear to have yet been determined.

Again, the subject of **bacteriology** probably has an intimate connection with the subject under consideration, and notwithstanding the infancy of this science there is already a vast accumulation of learning and research upon the subject. Indeed, with reference to any one of these subjects the field is sufficiently broad to occupy the best part of an ordinary life-time; so that, repeating what we have said before, we feel fully justified in this connection in confining ourselves to preliminary considerations, and not entering upon the subject of chemical analysis.

A poison is defined by Dr. Wormley, in his work upon the "Micro-Chemistry of Poison," as "any substance which, when taken into the body and either being *absorbed*, or by its *direct chemical action* upon the parts with which in contact, or when applied *externally* and *entering the circulation*, is capable of producing deleterious effects."

Dr. Taylor defines a poison as "a substance which, when absorbed into the blood, is capable of seriously affecting the health or of destroying life." This definition does not seem sufficiently comprehensive, and we prefer the definition of Dr. Wormley. The legal scope of the statutes against poisoning is not unfrequently enlarged by the use of other terms, such as "other destructive or noxious thing;" but the definition above quoted from Wormley is believed to be medically correct.

When the suspicion of poison's having been administered arises upon any occasion, the services of a medical practitioner or medical jurist may be called for under a variety of circumstances. A medical practitioner may be called upon to treat the party during life; to inspect the body after death; to testify upon the preliminary examination or upon the trial of the indictment, or upon both these occasions. In order to qualify him to act in these important relations, it is necessary that he should possess a most intimate acquaintance with the effect of a large majority of known poisons, and also with the course and termination of acute and chronic diseases generally.

Although the most satisfactory evidence of poisoning consists in the isolation of some particular poison from the tissues, or tissue secretions, or matter ejected from the body, there are some instances in which the medical probability of poisoning may, in conjunction with other circumstances of general evidence, raise so high a probability of poisoning that there can be little room for doubt, even though the particular poison has not been isolated.

It will of course be impossible in the scope of this manual to present anything like even an outline of the collateral sciences a knowledge of which is necessary to the full understanding of the subject under consideration. The student is referred for further particulars upon these important subjects to the general treatises upon Practice of Medicine, Materia Medica, Therapeutics, Pathology, etc.

In this connection, however, certain things may be touched upon which more particularly appertain to this subject, — such as the channels of entrance and mode of action of poisons, etc.

The channels of entrance of a poison into the human body may be thus enumerated: 1. The blood vessels, including wounds. 2. The skin and cellular membranes. 3. The air passages and lungs. 4. The stomach. 5. The intestines.

In this connection it may be well, perhaps, to specify **the channels of exit** by which poisons are eliminated or excreted from the body, and the organs and tissues in which they undergo an intermediate deposit. The channels of exit are: 1. The urine. 2. The bile. 3. The milk. 4. The saliva. 5. Mucous secretions. 6. Serous secretions. 7. The perspiratory fluid.

The organs or tissues where they undergo intermediate deposit are: 1. The liver. 2. The kidneys. 3. The spleen. 4. The heart. 5. The lungs. 6. The muscles. 7. The brain. 8. The fat. 9. The bones.

For further information upon these subjects, the student is referred to Wood's and Bartholow's excellent treatises upon Materia Medica and Therapeutics.

With respect to most poisons, their entrance into the blood, as the result of absorption or injection, is a condition necessary to their action. Strychnia or prussic acid, applied directly to the brain, spinal marrow, or nerves, produces no effect, or only a slight local action after some time; but when a portion of either of these

poisons is carried by absorption into the arterial capillary system symptoms of poisoning appear. Using,
however, the term "poison" in the broader sense in which
it is above defined, there appear to be some substances
in poisoning with which death is caused by local chemical changes; the mineral acids and caustic alkalies constitute the principal poisons producing this direct local
chemical effect.

Action of Poisons. — The greater portion of our
knowledge as respects the action of different poisons is
derived from empirical sources, such as observation of
the symptoms caused by them in the living, and the
morbid traces left by them on the dead body. A few
poisons destroy the tissues with which they come into
immediate contact; a larger number irritate and inflame
the part reached by them; a limited number act injuri-
ously on the brain or spinal cord, or both, or on both
the brain and the heart; while a large proportion of poisonous articles act either simultaneously or successively
both on the mucous surfaces and on the cerebro-spinal
system; on the former as irritants, on the latter either
as stimulants or sedatives. Some poisons have a direct
action on the organ or organs to which they are applied,
and an indirect action on the system generally, or on
some one or more of its remote organs. It is only in
the case of concentrated mineral acids, and a few of the
more powerful irritants, that the local action is sufficiently powerful to destroy life. What was once considered as the purely local effects of different poisons
is now known to be equally producible whether the
poison in question has been directly applied to the
organ affected, or has reached it through the circulation.

Arsenic, for instance, exerts its irritant action on the intestinal canal equally when swallowed, when applied to a wound, or when introduced into the veins by injection.

Certain poisons have an unquestionable affinity for particular viscera; thus, opium, chloroform, and chloral hydrate act on the cerebro-spinal system; strychnia on the spinal cord, etc. Such special actions are not, however, always witnessed on the organ for which they have a special affinity, and they are variably influenced by dose, mode of administration, constitutional peculiarities, and proclivity to morbid action, or the reverse, in the system. Certain poisons, without doubt, undergo changes within the body after their absorption and prior to their elimination. The mineral acids combine with alkalies, the vegetable acids are in some cases decomposed, in others eliminated unchanged, and in others enter into combination with the alkalies of the body.

The action of poisons is modified by a variety of circumstances, such as: 1. Quantity of dose. 2. Form in which the poison is administered. 3. Chemical action. 4. The tissues directly acted upon. 5. Habit. 6. Idiosyncrasy.

1. **Quantity of Dose.** — There are great differences among different drugs as respects the amount necessary to be taken to produce deleterious effects. There are few poisons which are not harmless in small doses, while many of them in moderate doses are medicinal. Some are so active that they can only be administered with safety in very small quantities, — such as strychnia, conia, veratria, aconitine, prussic acid, etc. Other poisons are injurious only when taken in very large doses, — such

as alum, sulphate of potash, cream of tartar, etc. One fiftieth of a grain of aconitine has produced alarming results to an adult; while an ounce of magnesium sulphate can generally be administered without danger, although in large quantities it has several times caused death.

In some instances the largeness of a dose of poison may save the patient by producing speedy and copious vomiting, leaving only irritability of the stomach for a few days.

Some poisons act on different parts of the system according to the amount of the dose.

2. **Form in which the Poison is Administered.** — The form in which the poison is administered has great influence upon its action. Some poisons require to be dissolved before they can produce their characteristic effect; others act most energetically in a state of vapor, such as chloroform or ether ; others, again, are weakened in their effect by admixture and dilution, although this rule is not universal.

3. **Chemical State.** — The chemical state of the poison materially affects its results; some substances cease to be poisonous when neutralized, while neutralization renders others more energetic; others, again, are not affected by neutralization provided the compound is soluble. Generally the effects of a base are little influenced by the acid with which it combines ; and the same rule generally holds true as to acids, such as prussic, oxalic, arsenious, and arsenic acids. Salts which are isomorphous are closely allied in action.

4. **The Tissues directly Acted upon.** — The tissue acted upon by the poison influences the results. The unbroken

skin is insensible to the action of most poisons unless when applied in a state of gas or vapor, or assisted
by friction. Many medicines may, however, be introduced into the system by inunction. The action of poisons introduced into the veins is very rapid and energetic,
as is also their action when brought into contact with
the air-cells of the lungs; next follow serous surfaces,
and lastly, mucous membranes.

Certain vegetable poisons which are very powerful
when applied directly to the wound, may be swallowed
with impunity.

5. **Habit.**—Habit has great influence in modifying the
effect of certain poisons. Its effect is most striking in
the case of opium and alcohol. Dr. Ogston, from whose
excellent work we abridge the principal part of this chapter, states that he has known a quart of laudanum being
consumed in a few weeks by an opium-eater, although
this was not done with ultimate impunity. He states
that habit has no effect in mitigating the action of inorganic poisons, and says that notwithstanding the alleged
facts which have been brought forward in proof of what
is termed the **tolerance** of increasing doses of certain of
these, — such as arsenic, tartar emetic, and sulphate of
copper, — the truth seems to be that beyond certain
limits, undoubtedly pretty wide in the case of arsenic,
the system in the use of these poisons becomes more
and more instead of less and less susceptible to the same
doses where their administration has been continued for
any length of time and in quantities at all considerable.
On the other hand, Dr. Wormley regards as authentic
the accounts of the Styrian arsenic-eaters published by
Dr. Roscoe, but states that the experience of most medi-

cal practitioners in the use of this drug does not accord with the results of this Styrian practice.

In certain pathological conditions of the body there is an increased susceptibility to the action of some drugs and a diminished sensibility to others. In delirium tremens, hydrophobia, tetanus, mania, or peritonitis, quantities of opium are beneficial which in health would be highly dangerous, if not fatal. On the other hand, where a predisposition to apoplexy exists, an ordinary dose of opium may cause death.

6. **Idiosyncrasy.** — Idiosyncrasy shows itself in an unusual susceptibility in some persons to the action of certain drugs in medicinal doses, or even in very small quantities. In some persons common articles of food or drink may produce symptoms of irritant poisoning. Dr. Ogston records one case in his practice in which opium in any shape or dose caused distressing nervous irritation; a second in which it led to erythema, and a third in which it brought on diarrhœa. Other interesting instances will be found narrated by the same author and by other writers upon this subject.

Evidence of Poisoning. — 1. **Suddenness of onset** of the symptoms in a healthy person soon after partaking some liquid or solid usually awakens suspicion of poisoning. This criterion is not usually a safe one, for the reason, as medical men well know, that there are many diseases whose onset is sudden and whose course is rapid; taken in connection, however, with other symptoms, these indications are of value. The time within which a poison takes effect after its administration, varies in different cases, — the action in some cases taking place either immediately or within a short time, and in others the symptoms being delayed an hour or even several hours. When several persons have partaken of the same liquid or solid and are all suddenly attacked in the same manner, this of course raises a much stronger presumption of poisoning.

2. **In cases of poisoning, the symptoms usually rapidly run their course,** their duration, however, being subject, even in the same drug, to considerable variation. A few poisons, notably prussic acid, very rapidly produce a fatal termination, although even in the case of hydrocyanic acid there are recorded instances in which death has not occurred until the lapse of several hours.

It should be remembered in this connection that there are certain diseases whose symptoms resemble irritant poisoning, such as cholera, inflammation of the stomach and bowels, and perforation of the stomach; and that the symptoms of other diseases may resemble narcotic poisoning, as in the case of apoplexy, inflammation of the brain, etc. In such cases the true cause of death may in some instances be discovered by the post-mortem inspection, while in others, there may be no characteristic morbid appearances.

Although, as above stated, symptoms resembling poisoning are not a safe criterion in many cases, still there are some cases in which alone they afford very strong evidence thereof; thus in the case of strychnia and oxalic acid it is conceded by Christison that there are no natural diseases which successfully simulate the effects of these poisons, and if it should happen that several persons who have partaken together of the same articles of food, drink, or medicine, have been simul-

taneously attacked with similar symptoms, the evidence
would amount almost to demonstration.

In some instances also very strong evidence may be
derived from the condition of the lips and mouth, the
impending suffocation from closure of the larynx, coldness
of the surface and depression of the heart's action, together
with acid stains upon the lips, mouth, and clothes; and
although the evidence from the symptoms except in a
few instances, may be short of actual proof of the administration of poison, they always constitute important
confirmatory evidence. It is to be remembered, however, as stated by Christison, that it does not follow because a poison has been given that it has been the
cause of death; hence, in every medico-legal inquiry
the cause of the first symptoms and the cause of the
death should be made two distinct questions.

3. **Evidence of Poisoning deducible from the Post-
mortem.** — The evidence from this source is either negative or positive. No reliance is to be placed on such
appearances in the dead body as unusual lividities or its
rapid decomposition; and it is believed that except in a
very few particular instances the morbid appearances
left by poison upon the corpse do not differ specifically
from those following natural diseases or some other
kinds of violent death. The entire absence of those
morbid appearances on the body which are indicative
of the action of poisons, even if conjoined with the presence of those which result from disease to an extent
sufficient to account for death from natural causes, will
not always suffice to negative the possibility of poison
having been administered. In such a case, however,
it will be very difficult if not impossible to secure a

conviction of poisoning. The only chance of the truth's being brought to light in such a case is where the symptoms and appearances produced by the disease are widely different from those caused by the poison alleged to have been taken.

In order properly to make a post-mortem inspection in a case of suspected poisoning, **the examiner should be well versed in pathology** and should be especially familiar with the numerous changes which may simulate the appearances due to the administration of poison during life; thus, while he should expect evidence of irritation, inflammation, or corrosion of parts of the alimentary canal after the ingestion of several of the irritant poisons, he must remember that all these may have arisen from natural causes. Thus the stomach of drunkards is sometimes found to be intensely red, or what is more common, reddening on exposure to the air. Appearances of superficial inflammation or even ulceration are sometimes encountered in the mouth, throat, and stomach of new-born infants, all of intrauterine origin. Again, decomposition may cause redness of the stomach; and in some cases of sudden death, as in executed criminals, the stomach has been found highly vascular where no symptoms of irritation or inflammation existed during life. Severe inflammation or ulceration in the mouth, throat or œsophagus observed in the dead body, are much more likely to have been the consequence of previous disease than of any irritant or corrosive poison. Corrosions of the stomach or smaller intestines from the action of poisons of this class will scarcely be confounded by a competent observer with ulcerations caused by disease.

Softening of the coats of the stomach is common both
to disease and to the action of corrosive poisons; such
softening has been known in a few instances of sudden
death to be caused by the solvent action of the gastric
juice on the coats of the stomach after death, or it may
present itself as an example of the disease termed gelat-
iniform softening; in both cases the tissues are completely disorganized and usually broken up into shreds
over a considerable portion of the organ. In such cases
the examiner should not attribute the appearance in
question to any corrosive poison unless the poison is
actually found in the stomach.

Rounded perforations of small size in certain unhealthy states of the stomach and smaller intestines
sometimes destroy life in a few hours, preceded by
symptoms not unlike the effect of irritant poisoning;
the symptoms where death occurs in such cases are
generally clearly due to peritonitis, while the edges of
the perforations are rounded, thickened, and of almost
cartilaginous hardness.

**In many of the deaths from narcotic and narcotico-
acrid poisoning, the inspection of the body affords no
evidence,** either the appearances being such as are
common to other cases of death, or there being no traces
whatever of morbid action in the body; this last may
happen even with some of the irritant poisons.

Putrefaction in the dead body may obliterate the
traces of a large number of those poisons which produce
marked changes in the fresh corpse.

Some poisons, however, leave traces of their morbid
action which are not easily obliterated; such as intense
phlogosis in the upper part of the alimentary canal from

certain of the irritants, and the destruction of its coats and even perforation of the tube by the more powerful escharotics, as well as the less marked effect on the skin, throat, or œsophagus by some of the mineral acids. It is, however, in connection with the symptoms and the general evidence that the morbid appearances after death furnish distinctive proof of poisoning.

Method of Making the Autopsy. — Inasmuch as the post-mortem will nearly always be conducted by a medical practitioner, and not by the chemist who makes the analysis, it is important that due care should be taken by the practitioner in making such examination, in order that the results of the chemical analysis may not be invalidated by the preceding negligence or ignorance of the party who made the post-mortem. The importance of this part of the examination can scarcely be over-estimated. The general method of making a medico-legal inspection has already been sufficiently treated in a preceding chapter; the attention of the reader will therefore here only be directed to some general considerations of importance in addition to those already given.

In inspecting the body of a person suspected to have been poisoned, it is of the utmost importance for the examiner before commencing his examination to be in possession of all the authentic information which can be procured regarding the previous history of the case. Such information is of importance at the inspection, as leading the inspector to attend particularly to the state of those organs which are most usually affected by the individual poison suspected to have been administered, and may prevent his overlooking points and circumstances to which otherwise his search and observation might not have extended.

Before proceeding to the inspection, the examiner should provide himself with ligatures, distilled water, and vessels for containing the stomach, the intestines, and any matters which may have been vomited; if this has not already been done, all suspected articles whether food, drink, or medicines, which may be found in the apartments of the deceased should be collected and carefully sealed up for future examination. If any substances have been spilt on the floor, a new or carefully washed sponge or clean cloth should be used to wipe them up, when they may be preserved in separate vessels. If the deceased has vomited, the vomited matters, especially those first ejected, should be preserved and their quantity, odor, color, and acid or alkaline reaction noted. If the vomiting has taken place on articles of dress or on the floor or furniture of the room, the stained portion of the clothing, sheet, or carpet should be cut out and preserved for analysis. If the vomiting occurred on a wooden floor, a portion of the wood may be scraped off or cut out; if upon a stone pavement, a clean piece of cloth or sponge soaked in distilled water should be used to remove any traces of the substance.

The vessels in which vomited matters were contained will sometimes present valuable evidence, since heavy mineral poisons fall to the bottom or adhere to the sides of the vessel.

When the vomited substance consists of animal matters liable to putrefaction it has been advised that they be kept in alcohol diluted with its own weight of water;

but the addition of any preservative fluid should be avoided, as it might complicate or embarrass future chemical analysis.

Urine and fœcal matters should also be preserved.

In making the external examination the procedure recommended in a previous chapter should be adopted. If any spots are found upon the skin, lips, or fingers, they should be removed with the scalpel for future analysis; the same thing should be done with the teeth when they present any appearances of having been stained or corroded.

In the examination of the interior of the body the chief attention should be directed to the alimentary canal; the œsophagus should not be disturbed in the examination of the trachea. The condition of the mouth and pharynx should be first noted before the organs in the chest are inspected. The left lung should be reversed upon the heart to expose the course of the œsophagus, and a ligature placed around it close to the diaphragm; two ligatures should then be placed round the duodenum near the pyloric end of the stomach and the duodenum divided between them to allow the removal of the stomach; the same thing should be done at the termination of the ilium and the smaller intestines removed. Finally, the extremity of the rectum being secured, the larger intestines are to be removed. The stomach and intestines should then be opened separately in clean glass vessels of known capacity; the contents of the separate portions of the alimentary canal, the state of their tissues and the alterations they have undergone should be carefully noted. The quantity, odor, acidity or alkalinity of the contents of the stomach, whether luminous or not in the dark, the presence or absence therein of crystalline matter, foreign substances, undigested food, and spirituous fluid, should be carefully noted. The appearance of the rugæ of the stomach and of their interspaces, particularly in the vicinity of the great cul-de-sac, should be noticed, as it is in this situation that the traces of poison and its effects are most frequently left. The seat of inflammation, if any, should be exactly specified, as also that of any softening or unusual coloration, ulceration, effusion of blood, corrosion, or perforation, if there should be any.

The same minute attention should be paid to the smaller and larger intestines and their contents. The parts of the intestines where morbid appearances are most frequently found in case of poisoning, are the duodenum, the upper part of the jejunum, the lower part of the ilium and the rectum. The comparative intensity of the appearances of irritation in different parts of the alimentary canal should be noticed, as it may throw much light on a suspected case of poisoning, and may sometimes obviate the necessity for any further proceedings: thus, if the stomach is sound and the intestines only inflamed the possibility of irritant poisoning may be pretty safely negatived.

After this examination of the stomach and intestines the organs are to be put into wide-mouthed glass vessels, each part by itself, and the vessels in which they were contained washed out with distilled water into the vessel appropriated to the morbid part, adding sufficient distilled water to cover the contained viscus.

When the poison has led to the perforation of the stomach and bowels the substance effused into the abdominal cavity should be carefully collected and separately preserved.

In a case of poisoning by arsenic and the greater number of the metallic poisons, the liver and the mass of the blood may require to be preserved. In cases of suspected poisoning with opium and its salts and all the vegetable alkaloids, besides the portion of the viscera most largely supplied with blood, portions of the blood and the whole of the urine should be kept.

The remaining organs of the abdomen must be inspected, particularly the spleen, the kidneys, and the rectum ; and in the female, the uterus and its appendages including the vagina.

No portion of the suspected articles should be wasted at the inspection or in preliminary trials or testing. In some instances it may be necessary to preserve, besides a portion of the liver and the blood, other parts of the body, — such as the kidneys, spleen, heart, brain, and portions of the muscles ; all the organs and the blood thus removed should be collected and preserved in separate clean glass vessels, great care being taken that none of the organs or substances thus removed at any time be brought in contact with any substance which might give rise to a suspicion of contamination. A sealed and signed label ought to be attached to each of the vessels before removal from the apartment. Great care should be taken to prevent the possibility of their being tampered with, and they should be retained in the sole possession of the inspector until delivered by him to a person duly authorized to receive the same.

It may be necessary occasionally to examine a dead body for poison, after a considerable period of interment.

Mere putrefaction should not be allowed to prevent an inspection necessary for medico-legal purposes. It has been shown by the experiments of Orfila and Lesucur that the decay of the body does not always render impossible the detection of the poison. They placed in a dead body, and allowed to remain there for some time, each of the following poisons, namely : sulphuric and nitric acid, arsenic, corrosive sublimate, tartar emetic, sugar of lead, protochloride of tin, sulphate of copper, verdigris, nitrate of silver, chloride of gold, acetate of morphia, chloride of brucia, acetate of strychnia, prussic acid, opium, and cantharides. They found that the acids become neutralized by the ammonia generated during the decay of the animal matter; that by the action of the animal matter the salts of mercury, antimony, copper, tin, gold, silver, and likewise the salts of the vegetable alkaloids, undergo chemical decomposition in consequence of which the bases become less soluble in water ; that acids may be detected after several years' interment, not always, however, in the free state ; that the bases of the decomposed metallic salts may also be found after interment for several years; that arsenic, opium, and cantharides undergo little change after a long interval of time, and are scarcely more difficult to discover in decayed than in recent animal mixtures ; but that hydrocyanic acid disappears very soon, so as to be indistinguishable after a few days.

4. **Evidence of Poisoning derivable from Chemical Analysis.**— We have already indicated our opinion that the chemical analysis in cases of suspected poisoning should be made by a competent chemist, and cannot, as a rule, be safely undertaken by a general medical practitioner.

For details as to the different methods of analysis the reader is referred to the treatises of Wormley, Taylor, and Christison.

In this connection the possibility should always be borne in mind that the poison may have been introduced into the body after death with the design to impute poisoning; or that although the poison was swallowed during life the fatal event may have arisen from natural causes; it should always be remembered that a large number of the best known poisons are constantly employed as medicinal agents.

5. **Experiments on animals** with suspected articles of food, drink, or medicine, were formerly much relied on as evidence of poisoning; now, however, this line of proof is properly objected to both as a waste of useful material, and on the ground that the effects of poisons on the lower animals are frequently different from those produced on man. For details see Ogston's "Medical Jurisprudence" and professed treatises on Therapeutics.

6. **With the moral aspects of the case** the medical jurist as such, has usually nothing to do; but in his capacity as a practitioner these aspects will not unfrequently be brought before his attention, and in the interests of justice should be carefully observed and noted by him. In many instances he may not only thereby materially aid the interests of justice, but he may sometimes thereby be enabled to protect his patient against further attempts at poisoning.

Treatment of Poisoning, generally. Before proceeding farther it may be well to insert at this place a few general principles as to the methods of treatment in cases

21

of poisoning. The details must of course be sought in larger works upon Therapeutics and Toxicology.

In treating a case of suspected poisoning there are three indications to be met: 1. The removal of the poison; 2. The counteracting of the primary effects of such portions of the poison as may have gotten beyond reach; and 3. The cure of the resulting disorders it may have occasioned.

The modes of removing the poison vary according to circumstances.

External Applications, etc. — Some poisons, such as the strong mineral acids and the caustic alkalies, act on the unbroken surface of the body, while the far greater number require to be introduced below the surface. Where the poison has been applied outwardly, careful ablution will in many instances be all that may be required to fulfil the first indication.

When, however, the poison has penetrated below the external surface, the following steps should be taken: A ligature should be applied between the wound or sore and the heart as far from the wound as possible, so as to arrest the circulation in the veins but not in the arteries; this will if adopted in time, prevent the absorption of the poison or arrest its further absorption, at least for a time. The next step to be taken is to wash the wound carefully from every trace of the poison, and to attempt the removal of so much of the poison as may adhere to the wound or have entered the circulation at the wounded part. Some authors advise the opening of a vein near the ligature, and the withdrawing of as much of the local blood as is necessary to insure the safety of the patient. Others recommend the manipulation of the soft parts around the wound so as by compression to squeeze out from it as much as possible of the blood it contains. A better procedure, and that generally recommended, is suction, either by the mouth or cupping glass; **dry cupping** is perhaps the best method and may be practised, if no better means are at hand, by burning a small piece of paper saturated with alcohol, in a small wine-glass or tumbler, and having removed the remains of the paper immediately placing the glass over the wounded part. Free scarification of the part will add to the efficiency of this operation.

Excision of the soft parts in the vicinity of the wound, or even the amputation of the part when the wound is situated in a finger or a limb, may sometimes be necessary.

Caustics or escharotics or actual cautery have been recommended.

When poisons have been applied to the unbroken surface or the entrance of the various mucous canals, so much of the noxious substance as possible should be speedily washed out; and in the case of the mineral acids and alkalies, the remainder neutralized with some appropriate reagent.

Where a volatile poison has been drawn into the lungs by inspiration it is rapidly absorbed and will usually be beyond reach before any attempt at its removal can be made; in such case the only resource within our power is the continuance of full and rapid inspirations and expirations to dilute and carry it off. Artificial respiration with or without the addition of vapor of ammonia or other appropriate drug, may be necessary.

In most instances in which the intervention of a practitioner is necessary, the poison will have been swallowed and the first most important aim in such case will be to procure its **complete and speedy removal from the stomach.** In some instances the poison will itself have provoked vomiting, which should in most instances be encouraged by the use of the mildest diluents. Where vomiting has not thus spontaneously occurred, or, if it has commenced, is not sufficiently free, emetics should be at once administered and repeated at suitable intervals till their full action is obtained. As a general rule, the milder vegetable emetics, such as ipecac or the preparations of squill, are best adapted for the irritant and many of the narcotico-acrid poisons; while sulphate of zinc, in a dose of five grains repeated every ten minutes until it has acted freely, will be found useful in the case of a narcotic poisoning. Common table-salt, in a dose of one or two tablespoonfuls dissolved in eighteen or twenty ounces of warm water, or, though more irritating, one or two teaspoonfuls of mustard in warm water, will be found useful in an emergency. Simple diluents may be given freely to assist the action of the above emetics and vomiting may be hastened or assisted by tickling the throat or compressing the epigastrium. There are cases, however, where it will be either unsafe to administer emetics or where if administered, they will fail to act; in the first category are patients predisposed to or who have already suffered from apoplexy or hæmoptysis.

In many cases the hypodermatic injection of from $\frac{1}{16}$ to $\frac{1}{12}$ of a grain of apomorphia will be found a most valuable agent in procuring the complete and speedy evacuation of the stomach. A strong recommendation in its favor is the fact that it can be used in cases where the patient cannot swallow, and in cases where, as in strychnia poisoning, the administration of an emetic or the use of a stomach-pump is difficult or impossible, by reason of convulsions. In many cases the stomach-pump may be safely used; its use, however, is sometimes difficult or impossible in the case of children, and as before mentioned, during the continuance of convulsions; while it may be altogether contra-indicated in some cases of corrosive poisoning. Where the patient's mouth cannot be opened, on account of spasms, it should be remembered that a small tube may be introduced into the stomach through the nostril. It may occasionally be necessary, as a means of fulfilling the first indication, to expel *per anum*, by the use of laxative medicines, portions of the poison which have passed from the stomach into the intestines. Castor oil is the cathartic usually chosen, on account of its mildness; when necessary its efficacy may be increased by addition to it of one or two drops of croton oil.

The second indication is the counteracting of the primary effects of such portions of the poison as have gotten beyond our reach; and the third indication is the cure of the resulting disorders the poison may have occasioned.

Inasmuch as it is seldom that the whole of the poison can at once be removed from the system by any means at our command, we may usually find it necessary to attempt to meet these second and third indications. To consider this subject at length or even partially would occupy more space than is at our disposal; we are obliged, therefore, to refer the student for particulars in this important subject to the excellent works of Drs. Wood and Bartholow on Therapeutics.

Feigned Poisoning, Imputed Poisoning, etc. — The imputation of the crime of poisoning by feigning or actually producing the symptoms and contriving that the poison shall be detected in the quarters where in actual cases it is usually sought for, has sometimes been attempted. It is very easy for an artful person to put poison into food and accuse another of having administered it, as well as to introduce poison into fæces or matters vomited. The possibility also of the introduction of a poisonous substance into the body after death, with a view of accusing an innocent person of the crime of poisoning, should also be borne in mind.

In cases of feigned poisoning, the absence of the characteristic symptoms of the poison alleged to have been taken, with other inconsistencies in the account of the party, will usually disclose the fraud. In such cases, in forming a diagnosis little reliance should be placed on the unsupported statement of the patient; the practitioner, while allowing the patient apparent credit for the truth of his statements and seemingly sympathizing with his fears, — real, imaginary or assumed, — should request him to give a full history of existing symptoms, their origin and progress, their relation in point of time to various meals, and the mode and vehicle in which the supposed poison was administered. The same course should be taken with attendants, witnesses or other interested in supporting the statement, either during his lifetime or after death in cases of suspected fatal poisoning. No unprofessional person can go through such an ordeal without bringing out many circumstances irreconcilable with the idea of poisoning generally, and still more the administration of a particular poison, unless it has in fact been administered.

In many cases it will be found that the party is not entirely free from some complaint. In imaginary poisoning there will often be disorder of the digestive organs, irritation of the alimentary canal, febrile or inflammatory symptoms, with or without delusion or hallucination showing that the mind is disordered. In suspected poisoning the same circumstances will have served as grounds for suspicion.

In feigned and imputed poisoning, if there be no illness to give a color to the imposture there will be a fictitious disorder produced. In all these cases, therefore, it should first be settled whether or not there be any actual departure from the state of health. This done, the additional evidence which the ejecta and dejecta (if any such are produced) are capable of affording, care being taken to ascertain that poison has not purposely been added to these, will usually be sufficient to determine the question. It may be necessary in order to settle the question definitely, to have the person and those about him secretly watched.

In a case of importance where active disease is found to exist, the medical jurist will not be justified in deciding positively that a poison has not been exhibited without a previous careful examination of the suspected articles, food or drink, and of the ejecta. The chemical investigation may sometimes lead to proof of the reality of an apparent case of feigned, suspected, or imaginary poisoning.

CHAPTER XXI.

INSANITY.[1]

THE compiler of this little volume has felt great embarrassment in the preparation of this chapter and the chapter upon Toxicology, on account of the wealth of material and the difficulty of determining what to omit; for with the limited space at his disposal nothing more than a mere outline can be attempted.

It is, perhaps, impossible to frame a perfect **definition** of insanity; if space permitted it would be a matter of great interest to collate and compare the different definitions, and to study their development. Quite a full compilation of the various definitions of insanity will be found in Dr. Hammond's work on that subject, who defines it as "a manifestation of disease of the brain, characterized by general or partial derangement of one or more faculties of the mind, and in which, while consciousness is not abolished, mental freedom is weakened, perverted, or destroyed:" insanity is regarded by this writer as, strictly speaking, "only a symptom of cerebral disease."

Dr. Spitzka gives the following comprehensive definition of insanity, which, he states, complies with the chief requirements of a practical definition, although

[1] The material for this chapter has with the author's consent been largely drawn from the manual of Dr. Spitzka upon Insanity, which every student should consult for further details.

laboring under the disadvantage of length: "Insanity is either the inability of the individual to correctly register and reproduce impressions (and conceptions based on these) in sufficient number and intensity to serve as guides to actions in harmony with the individual's age, circumstances, and surroundings, and limit himself to the registration as subjective realities of impressions transmitted by the peripheral organs of sensation; or the failure to properly co-ordinate such impressions and thereon to frame logical conclusions and actions; these inabilities and failures being in every instance considered as excluding the ordinary influence of sleep, trance, somnambulism, the common manifestation of the general neuroses, such as epilepsy, hysteria, and chorea, of febrile delirium, coma, acute intoxication, intense mental pre-occupation, and the ordinary immediate effects of nervous shocks and injuries." In another place the same author gives the following definition: "Insanity is a term applied to certain results of brain disease and brain defect which invalidate mental integrity. It is inaccurate to state that insanity is itself a disease. It is, strictly speaking, merely a symptom which may be due to many different morbid conditions, showing this one feature in common, — that they involve the organ of the mind."

Dr. Kiernan, of Chicago, has favored the author with the following definition: "Insanity is a morbid mental condition arising from brain disease, or disorder, or malformation, which perverts the mental relations of the individual to his surroundings, or to what from his birth, education, and circumstances, might be expected to be such surroundings."

Dr. Henry. M. Bannister of the Illinois Hospital for the Insane, at Kankakee, has also favored us with the following definition: "Insanity is a disease or defect of the brain, causing such disorder in the action of the mind as to affect the individual's conduct, put him out of relation to his surroundings, and render him liable to be dangerous or inconvenient to himself or others. Under insanity are not usually included the delirium of fever, or the direct effects of toxic agents, intoxicants, etc."

Dr. Bucknill, in his essay on "Unsoundness of Mind in relation to Criminal Acts," defines insanity as "a condition of mind in which a false action of conception or judgment, a defective power of the will, or an uncontrollable violence of the emotions and instincts, have separately or conjointly been produced by disease."

The same author has lately propounded a new definition which is intended to express the essential features of insanity from a legal point of view. He says: "Insanity is incapacitating weakness or derangement of mind caused by disease."

It does not appear to us that this definition adds greatly to our stock of knowledge upon this important subject; it only removes the difficulty one step further, without in the least aiding in the solution of the question as to what does amount to legal irresponsibility. But this subject will be referred to further on.

Dr. Tuke, in "Bucknill and Tuke's Manual of Psychological Medicine," regards insanity as "a condition in which the intellectual faculties, or the moral sentiments, or the animal propensities, — any one or all of them, — have their free action destroyed by disease, whether congenital or acquired." "He [the student] will not go far wrong," says Dr. Tuke, "if he regard insanity as a *disease of the brain* (idiopathic or sympathetic) *affecting the integrity of the mind, whether marked by intellectual or emotional disorder,* — such effects not being the mere symptom or immediate result of fever or poison."

It will be seen from the above definitions and others quoted in Dr. Hammond's work that there is a great difference of opinion as to what is a proper definition of insanity. We have quoted the above definitions for the reason that, excepting the proposed new definition of Dr. Bucknill, they seem to us to define the condition of insanity more clearly than those proposed by other authors. Some of these authors, it will be observed, regard insanity as a disease or defect of the brain, causing certain mental phenomena ; others apply the term to the phenomena themselves, or the mental condition which they represent, caused by a disease or defect of the brain ; or as Dr. Hammond expresses it, "insanity, strictly speaking, is only a symptom of cerebral disease." Dr. Spitzka likewise, as we have seen, characterizes insanity as merely a symptom.

In the present state of our knowledge, it seems to the writer better to apply the term "insanity" to the condition of mind evidenced by the mental phenomena observed, rather than to the disease of the brain causing such condition. This, however, is perhaps not a vital point, and any one of the three definitions above quoted of Dr. Bannister, Dr. Spitzka, or Dr. Hammond, will perhaps include the vast majority of cases.

It is probably more convenient than necessary that any attempt should be made to frame a definition of insanity which will include all, or the greater number of cases. A very superficial view of the growth of our knowledge upon this subject shows clearly the folly of attempting to frame a rigid definition. A definition which at the time it is made may appear to be perfect will, with the enlargement of our knowledge and experience, be found defective. As is well observed by Dr. Spitzka, the chief need of a definition is a medico-legal one, and we think that it may be stated that this need is not a pressing one; it is much better in our judgment, as in the case of fraud, to decide each case upon its individual facts, than to attempt to make it square with any rigid and, very likely, imperfect definition.

With reference to this subject, Dr. Spitzka observes: "On some occasions the question of defining what is called 'legal insanity' may be presented to the reader of these lines. When that question is asked he may safely challenge the questioner to show him a broken leg, or a case of small-pox in a hospital ward, which is not a broken leg or a case of small-pox in law; or show him a tumor, or case of softening of the brain, which is meningitis or sclerosis in law; or to define the condition under which any disease-symptom becomes an indication of health. When these conditions are complied with, and not till then, may physicians attempt to define 'insanity in law' as distinguishable from insanity in science; in the meantime he may rest contented with the dictum of one of the best legal authorities, 'that that cannot be sanity in law which is insanity in science, just as nothing can be a fact in science and a fiction in law at one and the same time.'"

The characteristic evidences of insanity are, by Dr. Spitzka in his manual, for practical purposes, divided into two groups: the somatic or physical indications, and the mental symptoms proper. The physical symptoms furnish evidence of the physical nature of the disorder, and are indications for medical treatment. It is, however, beyond the scope of this work to consider them in this place; for details upon this subject, see the treatises of Drs. Spitzka and Hammond. We may, however, in this connection mention, as frequent accompaniments of insanity, certain disturbances of sensibility; the facts that the functions of the intestinal canal in some forms of insanity show great disturbance, that the condition of the skin and its secretions are sometimes greatly altered; anomalies of appetite; motor and trophic disturbances, etc.

The mental symptoms proper, while furnishing hints for moral treatment, are principally valuable for general diagnostic and medico-legal purposes, and to these our attention will at present be confined.

The principal mental symptoms are the existence of hallucinations, illusions, delusions, incoherence, delirium, imperative conceptions, morbid propensities, a variety of emotional disturbances, and various disturbances of the memory, the consciousness, and the will. We do not mean to be understood as saying that all or any considerable number of these symptoms are to be expected in any one case, nor that any one may be considered as a criterion of the existence of insanity.

It may perhaps be well to define some of these symptoms, although little more can be attempted in this connection than a definition.

An illusion is defined by Dr. Hammond as being a false perception of a real sensorial impression. Illusions of all senses, but especially of sight and hearing, are found in insanity, and particularly in acute forms with delirium. Persistent presence of illusions, while not conclusive evidence of insanity, is evidence of brain disease, which, if it has not already caused insanity, may do so at any time.

An hallucination is defined by the same author as a false perception without any material basis, and therefore centric in its origin. These are always evidences of cerebral disease, and common phenomena of insanity.

A delusion exists where an illusion or hallucination is accepted and acted upon as real, and of the falsity of which the patient cannot be convinced. It is defined by Dr. Spitzka as a faulty belief out of which for the time being the subject cannot be reasoned by adequate methods.

The existence of delusions was formerly regarded as the true criterion of insanity. This rule was laid down by Sir John Nichol in the celebrated case of *Dew* vs. *Clarke;* but it cannot now be regarded as sound, inasmuch as there are undoubtedly cases of insanity in which there is no delusion. Where a delusion exists, however, it is clear evidence of insanity, provided it relates to a matter of fact, as distinguished from faith, and is contrary to the customary mode of thought of the individual, and held in opposition to such evidence as is logically opposed thereto. Where, however, a delusion, though based partly on faith, urges the subject to do some criminal act of violence, as, for instance, to sacrifice his son, — this would be considered an insane delusion.

Spitzka divides delusions into two classes, genuine and spurious; the former of which have been principally created by the patient himself, and the latter adopted from others. Genuine delusions are again divided by this author into systematized and unsystematized.

Incoherence needs no definition; it appears to be due to an inability to concentrate the attention, or to a lack of power in co-ordinating the different parts of the brain concerned in the formation and expression of ideas. It will generally be found some time or other in cases of acute mania, and is common in imbecility and in chronic insanity of any kind.

Delirium is defined by Dr. Hammond as being that condition in which there are illusions, hallucinations, delusion, and incoherence, together with a general excess of motility, and inability to sleep, and acceleration of pulse. It is commonly found in the first stage of acute mania, although it may exist at the beginning of any kind of insanity.

Morbid propensities and **imperative conceptions**, leading to what is called morbid impulse, are characteristics of those groups of insanity called pyromania, where there is a morbid impulse to commit arson, kleptomania, to steal, etc. The acts arising from these morbid impulses have been made by some writers the foundation of many different classes of insanity; these terms are perhaps convenient as a means of designating the leading symptoms of particular cases of insanity, but as a basis of classification are of little value.

Emotional disturbances exist almost universally in the insane. Spitzka lays it down as a cardinal canon of psychiatry that in insanity the moral feelings are usually more or less dulled or perverted.

The memory often suffers temporary disturbance in cases of acute and transitory forms, and in exacerbations of chronic forms. In terminal deterioration there is a progressive and permanent loss of memory.

Disturbances of consciousness may rarely result in a change in the sense of identity, and in the rare condition of double or alternating consciousness.

The will some time or other in the progress of the disease more or less disturbed in all cases of insanity.

Classification. — There are almost as many different systems of classification of the various forms of insanity as there are writers upon this subject. Classifications have been made in a variety of methods, depending on the standpoint from which the subject has been viewed. Insanity has thus been classified from the anatomical, physiological, etiological, psychological, pathological, and clinical or symptomatological standpoint. Considering the imperfect state of our knowledge upon the subject, the classification based upon the clinical or symptomatological characteristics of insanity would seem to be the most practical, and it is at the present time, perhaps, the system of classification adopted by the best authors. With advancing knowledge it is very likely that a new and more satisfactory classification may be made in some one of the other different methods above specified, but at present it does not seem possible.

We can in this connection do little more than reproduce the leading classifications of Krafft-Ebing, and of the three leading American writers upon this subject, — Dr. Spitzka, Dr. Hammond, and Dr. Ray.

The classification of Krafft-Ebing is as follows : —

A. Mental affections of the developed brain.
    I. Psychoncuroses.
        1. Primary curable conditions.
            a. Melancholia.
                α. Melancholia passiva.
                β. Melancholia attonita.
            b. Mania.
                α. Maniacal exaltation.
                β. Maniacal frenzy.
            c. Stupor.
        2. Secondary incurable states.
            a. Secondary monomania (secundaere verruecktheit).
            b. Terminal dementia.
                α. Dementia agitata.
                β. Dementia apathetica.
    II. Psychical degenerative states.
        a. Constitutional affective insanity (folie raissonante).
        b. Moral insanity.
        c. Primary monomania (primaere verruecktheit).
            α. With delusions.
                $\alpha\alpha$. Of a persecutory tinge.
                $\beta\beta$. Of an ambitious tinge.
            β. With imperative conceptions.
        d. Insanities transformed from the constitutional neuroses.
            α. Epileptic.
            β. Hysterical.
            γ. Hypochondriacal.
        e. Periodical insanity.
    III. Brain diseases, with predominating mental symptoms.
        a. Dementia paralytica.
        b. Lues cerebralis.
        c. Chronic alcoholism.
        d. Senile dementia.
        e. Acute delirium.
B. Mental results of arrested brain development : idiocy and cretinism.

Dr. Spitzka's classification is as follows : —

GROUP FIRST. — PURE INSANITIES.

SUB-GROUP A.

Simple Insanity, not essentially the manifestation of a constitutional neurotic condition.

First Class.

Not associated with demonstrable active organic changes of the brain.

I. DIVISION. Attacking the individual irrespective of the physiological periods.

a. *Order.* Of primary origin.

Sub-Order A. Characterized by a fundamental emotional disturbance.

Genus 1 : of a pleasurable and expansive character, — simple mania.

Genus 2 : of a painful character, — simple melancholia.

Genus 3 : of a pathetic character, — katatonia.

Genus 4 : of an explosive, transitory kind, — transitory frenzy.

Sub-Order B. Not characterized by a fundamental emotional disturbance.

Genus 5 : with simple impairment or abolition of mental energy, — stuporous insanity.

Genus 6 : with confusional delirium, — primary confusional insanity.

Genus 7 : with uncomplicated progressive mental impairment, — primary deterioration.

β *Order.* Of secondary origin.

Genus 8 : secondary confusional insanity.

Genus 9 : terminal dementia.

II. DIVISION. Attacking the individual in essential connection with the developmental or involutional periods. (A single order.)

Genus 10 : with senile involution, — senile dementia.

Genus 11 : with the period of puberty, — insanity of pubescence (hebephrenia).

Second Class.

Associated with demonstrable active organic changes of the brain. (Orders coincide with genera.)

Genus 12 : which are diffuse in distribution, primarily vaso-motor in origin, chronic in course, and destructive in their results, — paretic dementia.

Genus 13: having the specific luetic character, — syphilitic dementia.

Genus 14: of the kind ordinarily encountered by the neurologist, such as encephalo-malacia, hæmorrhage, neoplasms, meningitis, parasites, etc., — dementia from coarse brain disease.

Genus 15: which are primarily congestive in character and furibund in development, — delirium grave (acute delirium, *manie grave*).

Sub-Group B.

Constitutional Insanity, essentially the expression of a continuous neurotic condition.

Third Class.

Dependent on the great neuroses. (Orders and genera coincide.)

I. Division. The toxic neuroses.
  Genus 16: due to alcoholic abuse, — alcoholic insanity. (Analogous forms, such as those due to abuse of opium, the bromides, and chloral, need not be enumerated here, owing to their rarity.)

II. Division. The natural neuroses.
  Genus 17: the hysterical neurosis, — hysterical insanity.
  Genus 18: the epileptic neurosis, — epileptic insanity.

Fourth Class.

Independent of the great neuroses (representing a single order).

Genus 19: in periodical exacerbations, — periodical insanity.

Order: arrested development { Genus 20: idiocy and imbecility.
{ Genus 21: cretinism.

Genus 22: manifesting itself in primary dissociation of the mental elements or in a failure of the logical inhibitory power, or of both, — paranoia (monomania).

GROUP SECOND. — COMPLICATING INSANITIES.

These may be divided into the following main orders, which, as a general thing, are at the same time genera: Traumatic, Choreic, Postfebrile, Rheumatic, Gouty, Phthisical, Sympathetic, Pellagrous.

Dr. Hammond's classification is as follows : —

I. *Perceptional Insanities.* Insanities in which there are derangements of one or more of the perceptions.
  a. Illusions.
  b. Hallucinations.

II. *Intellectual Insanities.* Forms in which the chief manifestations of mental disorder relate to the intellect, being of the nature of false conceptions (delusions) or clearly abnormal conceptions.
  a. Intellectual monomania with exaltation.
  b. Intellectual monomania with depression.
  c. Chronic intellectual mania.
  d. Reasoning mania.
  e. Intellectual subjective morbid impulses.
  f. Intellectual objective morbid impulses.

III. *Emotional Insanities.* Forms in which the mental derangement is chiefly exhibited with regard to the emotions.
  a. Emotional monomania.
  b. Emotional morbid impulses.
  c. Simple melancholia.
  d. Melancholia with delirium.
  e. Melancholia with stupor.
  f. Hypochondriacal mania, or melancholia.
  g. Hysterical mania.
  h. Epidemic insanity.

IV. *Volitional Insanities.* Forms characterized by derangement of the will, either by its abnormal predominance or inertia.
  a. Volitional morbid impulses.
  b. Aboulomania (paralysis of the will).

V. *Compound Insanities.* Forms in which two or more categories of mental faculties are markedly involved.
  a. Acute mania.
  b. Periodical insanity.
  c. Hebephrenia.
  d. Circular insanity.
  e. Katatonia.
  f. Primary dementia.

g. Secondary dementia.
h. Senile dementia.
i. General paralysis.

VI. *Constitutional Insanities.* Forms which are the result of a pre-existing physiological or pathological condition, or of some specific morbid influence affecting the system.
    a. Epileptic insanity.
    b. Puerperal insanity.
    c. Pellagrous insanity.
    d. Choreic insanity, etc.

VII. *Arrest of Mental Development.*
    a. Idiocy.
    b. Cretinism.

Inasmuch as Dr. Ray's classification has long been before the profession, and has been adopted in whole or in part by many other writers, we reproduce it in this connection as follows : —

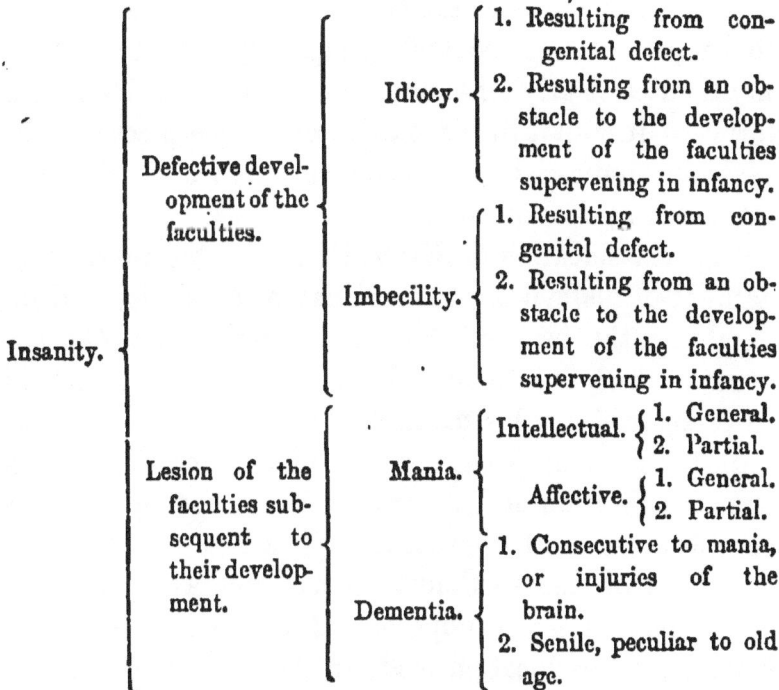

Insanity.

Defective development of the faculties.

Idiocy.
1. Resulting from congenital defect.
2. Resulting from an obstacle to the development of the faculties supervening in infancy.

Imbecility.
1. Resulting from congenital defect.
2. Resulting from an obstacle to the development of the faculties supervening in infancy.

Lesion of the faculties subsequent to their development.

Mania.
Intellectual. { 1. General. 2. Partial.
Affective. { 1. General. 2. Partial.

Dementia.
1. Consecutive to mania, or injuries of the brain.
2. Senile, peculiar to old age.

With respect to the subject of classification Dr. Ray observes that "it is not pretended that any classification can be rigorously correct, for such divisions have not been made by nature and cannot be observed in practice. Diseases are naturally associated into some general groups only, and if these be ascertained and brought into view, the great end of classification is accomplished. We shall often find them run into each other, and be puzzled to assign to a particular disease its proper place."

With reference to Dr. Ray's classification, which with some modifications is that adopted by Esquirol, it is to be observed that it is at present, as it seems to the writer, too narrow in its limits, and does not include many well-recognized forms of insanity.

Dr. Hammond's classification is not claimed by him to be perfect. Its several groups are not, so he says, in all cases clearly separated from each other. Particular forms of insanity, in his classification, are placed in the divisions indicated respectively by their chief manifestations of mental disorder.

The classification of Krafft-Ebing is objectionable on account of his making to some extent curability and incurability the basis of the same, as well as on other accounts which will be found well stated by Dr. Spitzka in Chapter XI. of his manual.

So far as we have been able to understand the subject, the classification of Dr. Spitzka, while somewhat more complex than other systems, appears to be the most scientific, and more exhaustive than any other classification which has been proposed. But as we have already stated, any classification made in the present state of our knowledge will probably be of comparatively temporary value, and must be modified with the advance of our knowledge of the subject.

We can in the further consideration of this subject do little more than define the principal sorts of insanity; and we desire here generally to express our indebtedness in this respect to the manual of Dr. Spitzka, to whose work we refer the reader for detailed descriptions of the different types of insanity.

Mania is a form of insanity characterized by an exalted emotional state, which is associated with a corresponding exaltation of other mental and nervous functions.

The typical condition of the maniac may be summarized in one phrase,— loosening of the inhibitions, or checks, both of organic and of mental life. The perceptions appear more acute; the associations are quick, so rapid, indeed, that the ease with which the patient forms new and extravagant combinations, and the readiness with which novel suggestions present themselves, impress the novice as manifestations of a naturally quick wit, or of a talented and original mind. Illusions are frequent, and hallucinations sometimes present. With the excitement in the sensorial, intellectual, and vegetative spheres, there develops a corresponding condition of the motor apparatus. Sometimes no furious stage is developed, and hallucinations are absent. The outbreak of mania is rarely, if ever, sudden or rapid; its duration varies, an average case having an initial stage of depression lasting about six weeks, a maniacal period of about three months, while the period of convalescence will occupy about a fortnight. Its prognosis is very favorable.

Melancholia is a form of insanity whose essential and characteristic feature is a depressed, that is, subjectively arising painful emotional state, which may be associated with a depression of other nervous functions. At its height it is the antithesis of mania. The common basis upon which the symptoms in all melancholiacs develop is the subjective painful emotion. Unsystematized delusions of a depressive nature are the most common symptoms. The delusions of the melancholiac may be modified by sensory disturbances, such as neuralgias, anæsthesias, or disordered smell and taste. Hallucinations are very common.

At some stages in the disorder the patient may become restless from fear, and spurious states of fury may thereby be developed. In mania the outbreak is the result of an expansive or angry emotion; while in melancholia, the outbreak best known as melancholic frenzy is the outcome of an anxious terror. This melancholic frenzy is of shorter duration than maniacal furor, and unlike the latter, it terminates suddenly.

The average duration of melancholia is from three to eight months, although it may last years. Its prognosis is less favorable than that of mania, although about six out of ten patients completely recover.

Katatonia is a form of insanity characterized by a pathetical emotional state, and verbigeration, combined with a condition of motor tension. According to Dr. Hammond, it is characterized by alternate periods, supervening with more or less regularity, of acute mania, melancholia, and epileptoid and cataleptoid states, with delusions of an exalted character and a tendency to dramatism. Its most striking phenomena are its cataleptoid periods. Its prognosis is favorable as regards life.

Transitory frenzy is a condition of impaired consciousness characterized by either an intense maniacal fury or a confused hallucinatory delirium, whose duration does not exceed a period of a day or thereabouts. This condition is designated by some observers as transitory mania, by others transitory melancholia; others have classed it among epileptic disorders. Dr. Spitzka prefers the term "transitory frenzy," in view of the specific feature of amnesia, and because this term commits to no doubtful hypothesis and best expresses the leading symptom of the disorder. It is a comparatively rare affection. Insolation, prolonged insomnia, exposure to extreme cold, violent emotional and intellectual strain, have frequently been determined to have been the exciting causes.

Stuporous insanity consists in the simple impairment or suspension of the mental energies, unmarked by any emotional or other perversion. At the height of this disorder the patient is in a state of immobility, and does nothing of his own initiative. Sensibility is impaired as much as mobility; the reflex acts are sometimes greatly impaired; the pupils are dilated, and react poorly; the pulse is tardy, small, and frequent; the temperature slightly lowered; the extremities are cold, while œdema of the feet is constantly, and of the hands and face sometimes, observed; mental activity shares in the depression, and abolition of other nervous and general somatic functions. The stuporous lunatic's recollection of the period of his illness is entirely destroyed.

This disease may run its course in a few weeks, but its usual duration is from one to three months. Its prognosis is highly favorable, nine-tenths of the patients recovering. This disorder is known in English and American asylums as acute or primary dementia.

Primary confusional insanity is a form of mental derangement characterized by incoherence and confusion of ideas, without any essential emotional disturbance or true dementia. This disorder is rare, and develops rapidly on a basis of cerebral exhaustion. Patients suffering from this disease after a rapid rise of their symptoms during a period of incubation rarely exceeding a few days, present hallucinations and delusions of a varied and contradictory character. Delusions of identity are very common. The speech of the patient is characteristic; sentences are left incomplete, and are entirely irrelevant as well as incoherent. Recovery is gradual; in only a small proportion of cases does the insanity remain and the patient become permanently deteriorated, his disorder then constituting a form of *chronic* confusional insanity.

Primary mental deterioration is an uncomplicated enfeeblement of the mind, occurring independently of the developmental and involutional periods. In most persons surviving their sixtieth year, a pronounced and general failure of the mental powers occurs at or after that period; this is the ordinary senile change and cannot be considered in all cases pathological; but where a similar deterioration anticipates the senile period, it can only be accounted for on a pathological basis. Such a decay of mind is observed in paretic, syphilitic, and organic dementia, and is also found to be the sequel of numerous other forms of insanity. In all these cases the mental failure is accompanied by active. symptoms characterizing the given variety of mental disorder.

In primary mental deterioration, however, progressive deterioration is chiefly limited to the higher mental faculties, and is the only notable indication of cerebral disturbance. The patient experiences a lack of energy both mental and physical; finds it difficult to go to sleep; there is an irritable condition of the brain manifesting itself in dreams; the patient becomes dyspeptic; and there may be signs of functional or organic heart disorder, or of the prodromal period of Bright's disease. At this stage a comparatively healthful mental state may be regained by rest and proper treatment; but if the exciting causes, which are principally excitement and mental strain, are kept in operation, actual dementia may follow. There are no delusions or morbid propensities, although there may be occasionally a suicidal tendency. Complete rest and proper tonic and moral treatment will check the disease at any but its later periods; but a complete restoration to the patient's former condition, Dr. Spitzka says he has never observed, even the most favorable cases revealing some permanent damage.

Secondary and Terminal Deterioration. — In certain cases of mania, melancholia, stuporous, and other primary forms of insanity, while death does not ensue, the patient does not recover, but a secondary and chronic psychosis develops from the primary disease. Terminal dementia is the ordinary conclusion of most chronic and uncured acute insanities. Numerous grades and varieties of this affection exist, and it is customary in characterizing these varieties to state the primary form which preceded them. The residua of the delusions of primary insanity may sometimes be detected.

Dementia must not be confounded with imbecility. While both dementia and imbecility imply a profound defect in the mental sphere, the former term should always be limited to acquired imbecility, and the latter to original feeblemindedness due to fœtal or infantile arrested development. The fundamental feature of terminal dementia is an acquired mental defect, which may vary from mere loss of memory, usually of recent events, or of the reasoning power, to a nearly complete extinction of mind, and a reduction of the patient to a mere vegetative existence.

Senile dementia[1] is a progressive and primary deterioration of the mind connected with the period of involution, but exceeding the ordinary extent of such involution to a pathological degree. It is characterized by an increased egotism or by penuriousness, enfeeblement of memory, prejudices formed on trivial or no ground, and frequently a profound moral deterioration. Unsystematized ambitious delusions are sometimes present, although the majority have depressive delusions, chiefly in respect to their property. Hallucinations and illusions may exist.

Should no other intercurrent illness cut short the course of the disease, bed-sores and colliquative diarrhœas, complicated sometimes with diseases of the bladder, terminate the patient's life. If the patient lives long enough, complete fatuity results, when he

---

[1] Senile dementia and senile insanity are not convertible terms.

may become voracious and filthy, to die finally with apoplectiform symptoms, or with those of a gradual and general paralysis.

Insanity of pubescence (hebephrenia) is characterized by mental enfeeblement marked by a silly disposition, following a primary period of depression which has the same tinge as, but without the depth of, that characterizing melancholia, and which coincides with or follows the period of puberty. It occurs between the fifteenth and twenty-second year, and begins with a period of sadness without, however, the depth of depressive emotion of melancholia, and in the midst of which there may be sudden outbursts of causeless laughter or silly jokes. After this preliminary period the patients may exhibit vague or blind propensities; have no settled aim, or may display stupid malice towards their surroundings. While there is no incoherence there is a peculiar tendency to verbosity and the use of long words or such as have an odd sound, etc. The intellect weakens progressively and the patient, who is usually a confirmed masturbator, gradually passes into a terminal dementia. Everything connected with the mental state of these patients appears shallow and even unreal.

The course of this form of insanity is protracted, and its prognosis exceedingly unfavorable. Many of the cases are still classed as primary dementia, particularly when the deterioration is very rapid. Where masturbation is a pronounced feature, some writers use the designation "insanity of masturbation."

Paretic Dementia. — The cases of insanity thus far described have been those whose essential characteristics are the mental symptoms proper. We now come to a second class of cases, in which there are demonstrable active organic changes of the brain. The first of these diseases to which we shall refer is paretic dementia, a disease presenting so many diverse symptoms and characteristics that we cannot hope, with the limited space at our disposal, to describe it with anything like completeness. It is characterized in its full development by a combination of mental and somatic deteriorations; the brain and the spinal cord of patients dying with this disease, show the results of long continued and often intense degenerative morbid processes, which many authors regard as of an inflammatory character. The brain itself is found to be wasted,—not uniformly so, however, but to a more marked degree in some, and to a less marked degree in other districts. A very frequent appearance in advanced paretic dementia is cystic degeneration of the cortex of the brain. But for details upon this subject reference is made to Chapter XIII. of Dr. Spitzka's manual.

The mental symptoms of this disease generally present the picture of unsystematized ambitious delusions, combined with progressive paresis and dementia; they may range, however, from atonic depression to the most furious delirium; from the construction of fanciful projects to extreme incoherency; from slight and almost undemonstrable mental impairment to the absolute extinction of higher mental life. The physical signs may vary from a slight disturbance of speech to gross paralysis, or may present themselves under the mask of a posterior spinal sclerosis (locomotor ataxia); of a disseminated organic disease, or of apoplectiform and epileptiform seizures. Among the individual signs there may be found almost any and every focal and general symptom known to the neurologist: paresis of various voluntary and involuntary muscles; anæsthesias, paræsthesias, hyperæsthesias, pains, and trophic disturbances; changes in the vascular tone; amblyopia, hemiopia, color-blindness, aphasia, choreiform and athetoid movements; progressive muscular atrophy; pseudohypertrophic, and bulbar paralysis. All these may be found co-existent with the mental disorder and, indeed, depending on the same morbid process.

The mental symptoms of the prodromal period are attributable to simple brain failure; the attention is not readily aroused; amnesia is noted from the beginning; there is more or less moral deterioration; there is a morbid irritability on slight provocation; simultaneously with this moral deterioration is a disturbance of the will and emotional nature.

In the active phase there are marked exacerbations of the physical signs; there are defects in the movement of the tongue and lips, so that the patient finds it difficult to pronounce, particularly the explosive and hissing sounds. A most characteristic feature is the association of other normally unnecessary movements with those of the lips and tongue; there is a tremor at the angles of the mouth; alternate dilation and contraction of the nostrils; corrugation of the eye-brows, etc., after which the word is thrown out precipitately. With these speech innervations all the finer motor co-ordinations seem to suffer. The depressive moods of the patient and complaints about head symptoms, if they existed, usually disappear about this time. Unsystematized delusious of grandeur appear, and are often joined with morbid projects and extravagant expenditures. Trophic disturbances and all the motor disturbances become more and more marked; bed-sores, furuncles, diarrhœas, pulmonary gangrene, etc., may supervene. Apoplectiform or epileptiform seizures sometimes mark the course of the disorder from the beginning, to which the patient may finally succumb.

The delusions of paretic dements although usually are not always expansive, but on the contrary may be depressive in their nature. Hallucinations and illusions sometimes exist.

There are three sorts of episodical attacks to which paretic dements are subject, which are among the most important signs of this disease, and which from their resemblance to maniacal delirium, epileptic fits, and apoplectic diseases, are called, respectively, maniacal, epileptiform, and apoplectiform, attacks of paretic dementia. For a particular description of these, the reader is referred to Dr. Spitzka's manual.

The prognosis of paretic dementia is very bad; there are, however, a few cases where there appears to have been a permanent recovery. The duration of the disease is commonly within three years, and in some cases it may terminate as soon as six months, although in other cases the disease may continue for six, or even ten years.

Syphilitic dementia bears a strong resemblance to paretic dementia, and to dementia from organic brain disease. In this disease demonstrable brain lesions, standing in constant relation to the symptoms, are found in a majority of cases. It is not always possible, clinically, to

make a sharp discrimination between syphilitic dementia and paretic dementia proper; syphilis plays a
prominent rôle in the etiology of the latter affection.
Delusions are not prominent and are rarely expansive.
After a prodromal period the course of the disease progresses very slowly towards a fatal termination.

Delirium grave is a comparatively rare form of derangement, approximating in many respects to maniacal
delirium. It has been variously termed, typhomania,
mania gravis, phrenitis, and acute delirium. It is
preceded and undoubtedly caused by profound nervous
or physical exhaustion and over-strain. Its mental
symptoms resemble the highest degrees of maniacal
furor and melancholic frenzy, but differ from these states
in their mode of development, in that the outbreak of
grave delirium is either sudden or preceded by a state
of impaired consciousness of a kind not found in mania
or melancholia proper. The ideation of grave delirium
is much more incoherent than that of frenzy, and is
usually the expression of an angry or frightened state.
At the outset the patient may still articulate sentences,
but his speech rapidly deteriorates and he is finally unable to pronounce syllables. The patients appear to
be afflicted with hallucinatory visions, of 'the day of
judgment, of conflagrations, of bloody scenes, or of those
connected with the exciting cause. There is great rest-
lessness; sometimes there are rhythmical motions; in
many cases there is absolute insomnia.

The second period of the disease is analogous to the
post-maniacal reaction following the outbreak of simple
mania; there is now extreme mental and physical depression. If the patient does not die in this condition,

he passes into a state resembling convalescence from
typhus, without the favorable termination of the latter.
The majority of patients afflicted with this disease die
in a delirious state within a few weeks. In those who
do not die within this period the excitement continues
without abatement four or five weeks, and the case ends
with fatal coma. Complete recovery rarely occurs: in
rare instances the patients emerge from the disease with
a slight mental defect ; in other cases paretic and terminal dementia supervene.

Chronic Alcoholic Insanity. — Insanity resulting from
alcoholic excesses ordinarily belongs to groups already described. The various states of drunkenness and delirium
tremens are not ordinarily considered insanity. There
are also various forms of dementia depending on organic
changes produced by alcohol in the brain and its membranes which pertain to the group of dementia from
organic disease. Just as epileptic and hysterical insanity may develop from the epileptic or hysterical
neuroses, so a special form of alcoholic insanity may
become engrafted on the alcoholic neurosis. It has distinct clinical characters, and is called chronic alcoholic
insanity.

On the chronic alcoholic constitution as a background,
with which most readers are sufficiently familiar, is
developed the psychosis of chronic alcoholic insanity.
After a brief prodromal period, marked by congestive
attacks and headaches, and under the influence of fright-
ful hallucinations, chiefly of visions, the patient becomes
the subject of delusions of persecutions, to which are
rarely added expansive delusions. The persecutory delusions of alcoholism relate to the sexual organs, sexual

relations, and to poisons. Many inebriates entertain also delusions of marital infidelity. Delirious exacerbations are likely to occur in consequence of the patient's morbid fear. Similar hallucinations to those found in acute alcoholic delirium may exist; the patient may see snakes, insects, dead bodies, mocking faces, etc. There is quite constantly more or less disturbance of the memory, and occasionally there is stupor.

The prognosis is very unfavorable, and there is a clear tendency to dementia.

Chronic Hysterical Insanity. — The forms assumed by hysteria are so manifold that anything like even a full enumeration of them would be impossible. So with chronic hysterical insanity. Its symptoms are various; the patients are changeable, emotional, fretful, careless, and superficial in their behavior and thoughts, extremely egotistical, and desirous of notoriety or sympathy, or both. To be the sufferer from an equally interesting, rare, and hopeless nervous disease is the ambition of some; to be considered the most abused woman on earth is the ambition of others. A patient with this hysterical character may develop psychoses quite analogous to those found in epileptic and alcoholic patients. There may be a transitory hysterical psychosis manifesting itself in deliria of fear; so we may have maniacal and melancholic states. The psychosis in hysteria may be protracted in duration, as in the alcoholic disorder. There is a tendency to simulate a theatrical behavior; there is an intensification of the hysterical character, to which is frequently added a silly mendacity. Sexual ideas are common, and manifested in two opposite extremes; hallucinations are frequent; in some patients obstinate mutism is observed, which by skilful cross-questioning will be found to be wilful.

A few cases of hysterical insanity and hystero-epilepsy in males have been described.

The prognosis of this form of insanity is unfavorable as to lasting recovery.

Epileptic Insanity. — Aside from epileptic dementia, a mental degeneration intimately dependent on the frequency of the convulsive attacks, and which may determine stupor, imbecility, or actual idiocy, according as these attacks begin later or earlier in life, — aside also from those attacks of furious madness replacing the convulsive attack, and which may be regarded as psychical equivalents of the convulsions, — there are forms of more or less protracted insanity following some individual epileptic attack, or breaking out in the interval, or finally extending over the entire interval, which are to be distinguished from the above forms.

Acute post-epileptic insanity may take the form of simple post-epileptic stupor, which may be complicated with dreamy deliria or with illusional or hallucinatory confusion and verbigeration; or post-epileptic morbid conditions of fear or fright, either simple or complicated with *delire raissonante* or great excitement; or post-epileptic maniacal moria, — a rare form, which closely simulates ordinary acute mania. There are also cases of chronic protracted epileptic insanity closely related to the post-epileptic forms.

Dr. Spitzka suggests the following chronological classification of epileptic insanity: —

1. The epileptic psychical equivalent which replaces the convulsive attack.

2. The acute post-epileptic insanity which almost immediately follows the convulsive attack.

3. The pre-epileptic insanity which precedes the outbreak of the convulsive attack or its equivalent, and increases up to the moment when the paroxysm explodes.

4. The purely intervallary epileptic insanity which, neither immediately following nor preceding a paroxysm, occurs in the interval between such.

Dr. Spitzka says it is possible for all these forms to occur together, and in addition there is very apt to be a background of protracted epileptic dementia; it is only where epilepsy is recent that the above forms are found in an unmixed state.

The immediate prognosis of epileptic insanity is favorable as regards the more acute explosions; the protracted forms are sometimes recovered from, but mental enfeeblement is more likely to ensue than in the former case.

Periodical Insanity is characterized by the recurrence of mental disorder at more or less regular intervals, the attacks being separated by periods during which the patient presents a state of apparent mental soundness. An important characteristic of periodical insanity is the similarity of the manifestations of the different attacks in the same patient for long periods. The intervals between these attacks are not always entirely lucid, but rather sub-lucid. It may take the form of periodical mania or periodical melancholia.

Periodical insanity does not always manifest itself under the guise of a single form of derangement. There is a subdivision, known as **Circular Insanity**, character-

ized by the alternation of mania and melancholia in
regularly recurring cycles, the order of which varies in
different patients.   As a rule, the mania and melancholia correspond to each other in intensity, and, generally,
the shorter the cycle the more intense the symptoms
and the better the prognosis.   Circular insanity generally begins at or about the age of puberty, and,
like other periodical insanities, is more frequent with
females than with males, is intractable to treatment,
and while not ordinarily leading to dementia, some
mental deterioration sooner or later is manifested.

The States of Arrested Development. — Under this
head are included the conditions known as **idiocy, imbecility, and cretinism.**   They are usually divided into
three grades: a subject deprived of all higher mental
power, and who is unable to acquire the simplest accomplishments is termed an **idiot**; one who is capable
of acquiring the simplest accomplishments, but who is
unable to exercise the reasoning power beyond the extent to which a child is capable, is termed an **imbecile**;
finally, there is a large class of subjects who are defective in judgment, and in whom this defect is of similar origin to — though not as intense as — that of the
imbecile and idiot.   These classes are not separated by
any distinct margin of demarcation, but shade indefinitely from one class to the other.

In **idiocy** there is usually, besides the mental defect,
some deficiency in the peripheral organs or their functions.   Many idiots are deaf or mute, or both; some
are blind, and anæsthesia as well as anosmia has been
observed.   They learn to walk late or not at all; the
skeleton is usually poorly developed; rachitis is com-

mon, and the somatic functions are generally imperfectly performed; the sexual organs particularly are found to be rudimentary or deformed, and the sexual function is usually in abeyance in idiots, although this is not always the case. The imitative tendencies in idiots are often very strong; in imbeciles they may be utilized to make good artisans. In the lowest forms of idiocy speech may be altogether absent or limited to a few inarticulate sounds; in others, a few words or short sentences may be acquired. Idiots rarely reach maturity.

The mental state of the **imbecile** has been very well expressed by the statement that "those mental coordinations acquired in the course of a higher civilization have not been formed in him." While defective as to reasoning capacity, his emotional state may present every analogy to that of healthy persons or approximate that of other forms of insanity. Moral defect is a prominent feature of some cases. In Dr. Spitzka's opinion the term "moral insanity" of authors should be limited to this class of subjects, and a much better term to use would, in his opinion, be " moral imbecility."

Morbid projects, imperative impulses, and morbid egotism are common in some imbeciles, and in such cases it may be difficult to decide whether they pertain to the group of imbecility or of original monomania.

Both imbecility and idiocy are sometimes marked by other disturbances of the nervous functions than those comprised in the mind. Epilepsy is common; spastic symptoms, contractures, strabismus, peculiar speech-defects, and stuttering sometimes exist. The condition of idiocy and imbecility is usually a stable one,

although occasionally progressive deterioration is caused
by epileptic fits.

The mental phenomena of cretinic idiocy are like
those of ordinary idiocy. For further description of
this subject the reader is referred to Dr. Spitzka's Work,
and to other treatises on nervous and mental diseases.

Paranoia (Monomania) — Monomania, as it has hitherto
usually been called, or as Dr. Spitzka now terms it,
"paranoia," is a chronic form of insanity based on an acquired or transmitted neuro-degenerative taint, and manifesting itself in anomalies of the conceptional sphere;
which, while they do not destructively involve the
entire mental mechanism, dominate it.

The symptoms of classical paranoia may be numerous
or varied, or they may be few or limited in range. In
some patients, usually encountered without asylums, a
single imperative conception or impulse, or a delusive
suspicion which may never become organized into an
insane belief, may be the sole mental symptom. De-
lusional monomania is the most frequent form of this
disorder. The delusions of this form of monomania
are alone sufficient to characterize the disorder, and
when found serve to establish the diagnosis. They are
of the systematized variety. Delusions of persecution
are the most common ones; hallucinations, particularly
of hearing, are very common. Their beliefs are almost
as numerous as the patients, but are all characterized
by the feature that the occurrences in the outer world
are anxiously examined by the patient with a view of
tracing their connection with himself. Sometimes, as
the result of inward reflection and reasoning, there is
a rapid transformation of the delusions of persecution

into those of aggrandizement. The varieties are numerous. (For details see the larger treatises upon Insanity and Medical Jurisprudence.)

The outbreak of the disorder usually coincides with some one of the physiological periods, such as puberty, the second climacteric, pregnancy, and the puerperal state. It is sometimes precipitated by sexual excesses, and occasionally by visceral disease and fevers.

The prognosis is very unfavorable. The disease, when not cured, remains stationary for years. Mental' deterioration, however, does not proceed rapidly,* and never reaches the degree of chronic confusional insanity or terminal dementia unless there is some intercurrent disease.

Simulation of Insanity. — Insanity may be simulated by ignorant persons, or by persons who have had more or less opportunity for observing or studying the disease. In the former case detection is easy; in the latter it is more difficult, and there are recorded instances where competent observers have been deceived by simulators. (See instances narrated by Dr. Spitzka.)

A common test with some persons is that the simulator does not repudiate his insanity as does the truly insane person. Aside from the fact that the insane do sometimes recognize their condition and exceptionally admit it, this criterion is at present entirely worthless, for the reason that many simulators now know that it is by some considered as a test, and govern their actions accordingly.

Usually the observer's attention will be directed to the possibility of simulation by some inconsistency in the clinical picture presented by the subject. Clearly

marked cases of different kinds of insanity before described, are very difficult to feign correctly in every
feature, and the simulator, through ignorance, will not
unfrequently confound different types of disease and
thus disclose the fraud. But there are some obscure
and mixed groups not easily recognized, in which a skilful simulator may succeed in imposing upon even a
competent observer.

The existence or non-existence of a motive for simulation will often throw much light upon the subject.
The simulator, by his adoption of the popular idea of
insanity, that the insane are either raving and incoherent, at all times and on all points, or in a condition
of fatuity, will thereby frequently reveal the fact of
simulation. A knowledge of the clinical features of
the different sorts of insanity will usually enable one
to expose such simulation. As a rule, the simulator, in
those quiet periods following upon his artificial excitement, which are the expression of inability to maintain
the exacting efforts of simulation, does not recognize his
friends or surroundings, or recollect anything that occurred about that period of time in which he has a
motive for making people believe he was irresponsible.
The true maniac, however, will be lucid in these very
periods, and will recollect his family and his friends perfectly well; and if he has committed a crime, while he
may be acute enough to desire to conceal his recollection
of it, he will not, if the examination is led up to the
period of its commission, gradually claim to forget real
circumstances occurring before and after it, as does the
simulator. Aside from toxic, hysterical, and other transitory morbid conditions, it is only in epileptic mania

and in paretic dementia that such amnesia really occurs; but here the physical signs or the history, or both, will afford unmistakable evidence of these affections if they really exist. The simulator will also err generally in allowing his feigned disorder to explode and to recede too rapidly.

Another characteristic feature with many simulators is the intensification of their symptoms when under examination. The simulator also labors under the mistake that the insane do not reason; too great a degree of incoherence in a delusion justifies a doubt of its genuineness. Persons feigning a quiet sort of insanity usually attempt to imitate dementia. The simulator of imbecility or dementia either talks more confusedly than harmonizes with the thread of reason he unwarily exhibits, or he talks less confusedly than he should in the utter absence of a connecting bond in his thoughts; in short, he does not balance the defects in ideation and in their expression properly.

The absence of insomnia and impaired digestion in the acute psychoses is exceptional in real insanity, and is, in that respect, a ground of suspicion. The simulator's task is rendered difficult whenever he is kept under continued observation. The best actor may fail to adhere to his assumed character for days and nights in succession.

Dr. Spitzka mentions the following special signs as justifying the suspicion of simulation : —

1. Studious efforts to avoid looking at the physician upon his entrance.

2. Extravagantly absurd answers to simple questions.

3. Taking a long time to answer questions, and hesitation in answering.

4. Furtively glancing when he supposes himself unwatched to see if any one is approaching to necessitate his being on guard.

5. A person feigning epileptic and somnambulistic states may recollect perfectly his feigned acts and expressions, and carry them into his quasi-lucid period.

6. Rhythmical movements are made by certain simulators which may have no analogy in insanity or are out of harmony with the form of mental disturbance assumed.

7. Simulators complain much more about odd and painful sensations in the head than the insane usually do.

8. A clumsy simulator may say, "I have the delusion that I am lost," etc., or, "I have hallucinations of faces," etc. Such a person can be readily exposed to be a simulator on other grounds, but the feature here mentioned alone suggests simulation. A true lunatic may admit he has delusions or hallucinations, especially when examined for the purpose of being committed to an asylum; but when he does so he affects to admit that he *imagines* those things; but a real lunatic never gives them names, showing that he recognizes their abnormal nature: he *is* lost; he *is* pursued by the devil; he *hears* voices, and he *sees* faces.

9. It is suspicious if insanity appears immediately after a crime, or after an arrest, or sentence, where its previous existence can be disproved.

Among devices which may be legitimately resorted

to to expose simulation, Dr. Spitzka mentions the following:—

1. When examining the patient let the speaker remark in an undertone to a by-stander, that if such and such a sign were present he would know in which ward to put him, or know in which form of insanity to classify the subject.

2. While being examined as to his general sensibility the simulator may believe that anæsthesia is a desirable part of the clinical picture; he will wince when probed with a pin unexpectedly, but will remain immoble when pricked after being warned.

3. When a simulator is accused of shamming he may either turn away from the examiner or suddenly lapse into stupor or undergo some other unnatural change of the symptoms. A real lunatic will either act as a sane person under those circumstances, or, as in apathetic states, show no change whatever.

4. A simulator, if transferred from one ward to another of an asylum, will imitate the different forms of insanity he sees there. Imitations may occur in real insanity, but it is limited to delusive conceptions which are accepted by the weak-minded lunatics from more intelligent ones,—in what the French call *folie communiquée* and *folie à deux.*

Simulation should not be directly charged until all other means have been exhausted. Anæsthesia, by the use of ether, may sometimes be of value; and the application of the Faradic wire brush may sometimes expose the simulation. It should be remembered in this connection that the insane sometimes simulate a different form of insanity from that which they actually have; this combination of real and feigned disease is by no means rare. See Dr. Spitzka's Work, and an article upon the "Simulation of Insanity by the Insane," by Dr. Kiernan, in the "Alienist and Neurologist," April, 1882.

The legal effect of Insanity may come before courts of law for decision in a variety of cases:—

1. Where insanity is pleaded as a defence to an indictment for crime.

2. In civil causes where the insanity is alleged in order to supersede a person in the management of his affairs, or where it is alleged for the purpose of avoiding a contract or will.

3. In either civil or criminal causes when it is objected to a witness that he was insane at the time of the occurrence of the events of which he is to testify, or that he has had an attack of insanity between those events and the trial,—which objection is by Dr. Ogston said to be valid in Scottish practice, although it is not either in England or in this country.

1. The attempts of the courts to fix upon a criterion by which to settle **legal responsibility for crime in** cases of alleged insanity have been numerous, but have not thus far met with success. By most of the courts the law is still laid down in accordance with the doctrine of *McNaghten's Case*, which will be found reported in vol. 10 of Clark & Finnelly's "Reports of Cases in the House of Lords," p. 200. In this case McNaghten was indicted for the murder of Drummond by shooting on Jan. 20, 1843, and the verdict was "Not guilty, on the ground of insanity." This verdict and the question of the nature and extent of the unsoundness of mind

which excuse the commission of a felony of this sort
having been made the subject of debate in the House
of Lords, the opinion of the judges on the law governing such cases was required upon a series of questions ;
to which they answered in substance that the responsibility of a person alleged to be insane, and who is accused of crime, depends upon whether or not the accused
at the time had sufficient capacity to know the nature
and quality of the act he was doing, and whether what
he was doing was right or wrong. As the case is one
frequently referred to, we have thought it proper to
state the questions and answers thereto at length in a
note.[1]

[1] Tindall, C. J. : " The first question proposed by your Lordships is
this: 'What is the law respecting alleged crimes committed by persons afflicted with insane delusion in respect of one or more particular
subjects or persons ? — as, for instance, where at the time of the commission of the alleged crime the accused knew he was acting contrary
to law, but did the act complained of with a view, under the influence
of insane delusion, of redressing or revenging some supposed grievance
or injury, or of producing some supposed public benefit.'

" In answer to which question, assuming that your Lordships' inquiries are confined to those persons who labor under such partial
delusions only, and are not in other respects insane, we are of the opinion that, notwithstanding the party accused did the act complained of
with a view, under the influence of insane delusion, of redressing or revenging some supposed grievance or injury, or of producing some public benefit, he is nevertheless punishable according to the nature of the
crime committed, if he knew at the time of committing such crime that
he was acting contrary to law ; by which expression we understand
your Lordships to mean the law of the land.

" Your Lordships are pleased to inquire of us, secondly, 'What are
the proper questions to be submitted to the jury, where a person alleged to be afflicted with insane delusion respecting one or more particular subjects or persons is charged with the commission of a crime
(murder, for example), and insanity is set up as a defence ?' And

The criterion laid down in this case is, to say the least, very unsatisfactory. Although without doubt a person who is by reason of mental disease unable to comprehend the difference between right and wrong in a particular instance ought not to be punished as for a crime, the criterion under consideration is entirely of too limited application, and if strictly applied must lead to the conviction and punishment of many persons clearly irresponsible.

The limited space at our disposal will not permit the full discussion of this important subject; the reader is accordingly referred to the very clear and conclusive argument upon the subject by Dr. Ray, in the first

thirdly, 'In what terms ought the question to be left to the jury as to the prisoner's state of mind at the time when the act was committed?' And as these two questions appear to us to be more conveniently answered together, we have to submit our opinion to be that the jurors ought to be told in all cases that every man is to be presumed to be sane, and to possess a sufficient degree of reason to be responsible for his crimes, until the contrary be proved to their satisfaction; and that to establish a defence on the ground of insanity, it must be clearly proved that at the time of the committing of the act the party accused was laboring under such a defect of reason, from disease of the mind, as not to know the nature and quality of the act he was doing; or if he did know it, that he did not know he was doing what was wrong. The mode of putting the latter part of the question to the jury on these occasions has generally been, whether the accused at the time of doing the act knew the difference between right and wrong; which mode, though rarely, if ever, leading to any mistake with the jury, is not, as we conceive, so accurate when put generally and in the abstract as when put with reference to the party's knowledge of right and wrong in respect to the very act with which he is charged. If the question were to be put as to the knowledge of the accused solely and exclusively with reference to the law of the land, it might tend to confound the. jury, by inducing them to believe that an actual knowledge of the law of the land was essential in order to lead to a conviction; whereas the chapter of his work upon the Medical Jurisprudence of Insanity.

This doctrine has not always been approved by legal authors and courts. Upon this subject Mr. Bishop, in his work upon Criminal Law, vol. i. sect. 381, says that "in the criminal law insanity is any defect, weakness, or disease of the mind rendering it incapable of entertaining the criminal intent, which constitutes one of the elements in every crime."

In the further consideration of the subject he observes that a legal test for insanity has never been found, because it does not exist; that the question

law is administered upon the principle that every one must be taken conclusively to know it, without proof that he does know it. If the accused was conscious that the act was one which he ought not to do, and if that act was at the same time contrary to the law of the land, he is punishable; and the usual course therefore has been to leave the question to the jury whether the party accused had a sufficient degree of reason to know that he was doing an act that was wrong; and this course we think is correct, accompanied with such observations and explanations as the circumstances of each particular case may require.

"The fourth question which your Lordships have proposed to us is this: 'If a person under an insane delusion as to existing facts, commits an offence in consequence thereof, is he thereby excused?' To which question the answer must, of course, depend on the nature of the delusion; but making the same assumption as we did before, namely, that he labors under such partial delusion only, and is not in other respects insane, we think he must be considered in the same situation as to responsibility as if the facts with respect to which the delusion exists were real. For example, if under the influence of his delusion he supposes another man to be in the act of attempting to take away his life, and he kills that man, as he supposes in self-defence, he would be exempt from punishment. If his delusion was that the deceased had inflicted a serious injury to his character and fortune, and he killed him in revenge for such supposed injury, he would be liable to punishment."

whether in a particular instance the act alleged to be
a crime proceeded from a sane or insane mind, is a pure
question of fact for the jury, and not of law for the court;
as, for example, it is a question of fact for the jury, and
not of law for the court, whether there is such a disease
as dipsomania, and whether the act in question was occasioned by this disease, or was the act of a sound mind.[1]

This seems a much more rational method of procedure than to attempt to define any rigid test. Where
each case is determined upon its own circumstances by
the evidence of competent experts before a jury, in the
same manner as other disputed questions of fact are determined, there will be comparatively little danger of
going astray, although it is possible that there might be
improvements in the method of securing the opinion of
such experts.

For the further consideration of this subject the
reader is referred to the preliminary chapter of Ray's
" Medical Jurisprudence," and to chap. 26, vol. i.
Bishop's "Criminal Law."

2. The procedure upon inquiries as to the idiocy or
lunacy of any person having estate, real or personal,
which is likely to be dissipated so as to expose himself
or family to want or suffering, is regulated by statute
in most of the States, to which statutes the student is
referred. The question in such cases as to the inability
of a person to manage his affairs, on the ground of insanity and consequent incapacity, is one of fact to be

[1] See 1 Bishop Crim. Law, § 383; *The State* vs. *Pike*, 49 N. H.
399; *Bradley* vs. *The State*, 31 Ind. 492; *The State* vs. *Jones*, 50
N. H. 369; *The State* vs. *Johnson*, 40 Conn. 136; *Stevens* vs. *The
State*, 31 Ind. 485; 4 Law Review, 236.

determined by the circumstances of each particular case. The principles hereinafter stated will throw some light upon the question of capacity in such cases.

ʿ If physicians who have certified to the insanity of a person have not made the inquiry and examination which the statute requires, or if their evidence and certificate in any respect of form or substance are not sufficient to justify a commitment to an asylum, the authorities should not commit; and if they do, it is not the fault of the physicians, provided the latter have stated facts and opinions truly, and have acted with due professional care and skill.

If a medical man takes upon himself the responsibility of imprisoning a person on the ground of insanity on mere statements made to him by others, he will be liable to an action unless he can show that the party imprisoned was insane at the time. A medical man or any other person may, however, justify an assault where it is committed for the purpose of putting restraint on a dangerous lunatic in such a state that he is likely to do mischief to some one ; this restraint, however, cannot be continued indefinitely without due process of law, but only so far as may be necessary to prevent damage and to have his case properly passed upon by legally constituted authorities.

It is well settled that insanity may be pleaded by an insane person in avoidance of his contracts made while insane. The degree of unsoundness of mind required to incapacitate a person from contracting may be stated to be such a condition of insanity or idiocy as from its character or intensity disables him from understanding the nature and effects of his acts, and therefore disqualifies him for transacting business and managing his property. It is well settled that in the absence of fraud, imposition, or undue influence, mere weakness or feebleness of understanding short of this is insufficient.

One who seeks to set aside a contract upon the ground of insanity alone, general or partial, must show that it was the offspring of mental disease.. Thus monomania, in no way connected with the subject of the contract, will not invalidate it. The unsoundness of mind required to vitiate a contract must also exist at the time of making such contract.

The contract of a person *non compos mentis* is voidable only, and not void; and hence on his restoration to reason may be ratified or avoided by him. After a judicial finding of the fact of insanity, however, the deeds and other contracts of persons *non compos mentis* are held to be void; though in some cases this seems to be the result of statutory provisions. The marriage of a person *non compos mentis* is, according to the better opinion, void.

As to the executed contracts of insane persons, the case of *Molton* vs. *Camroux*, 2 Exchequer Reports, 487; s. c. 4 id. 17; Ewell's Leading Cases, 614, lays down the rule that where a person apparently of sound mind, and not known to be otherwise, enters into a contract which is fair and *bona fide*, and which is executed and completed, and the property the subject matter of the contract cannot be restored so as to put the parties *in statu quo*, such contract cannot afterwards be set aside either by the alleged lunatic or by those who represent him. The rule of this case has been adopted in New Hampshire, Iowa,

New York, Pennsylvania, New Jersey, South Carolina, Indiana, and perhaps other States ; and on principles of public policy must ultimately prevail everywhere.

Although a lunatic has not a general capacity to enter into contracts, it is well settled that he may bind himself by implied contracts for necessaries suitable to his condition in life. It is well settled, also, that idiocy or lunacy is no defence to an action for a tort, that is, a wrong not connected with contract.

A contract entered into by a lunatic during a lucid interval will also bind him. By the term "lucid interval" is to be understood a condition in which there is a cessation of the symptoms of mental aberration, and a restoration to reason occurring between two paroxysms of insanity; such a total cessation of the symptoms of mental aberration and such a complete restoration to reason, according to Dr. Hammond, probably does not exist except in recurrent and epileptic forms of insanity, and in certain varieties of monomania, and of morbid impulse. He well states that the idea of a lucid interval being a temporary cure is now confined to the writings of those whose notions of the disease have been derived from books rather than the wards of hospitals. Like most other diseases, insanity is subject to remissions more or less complete, and there is no more propriety in regarding them as recoveries than there would be in considering the interval between the paroxysms of a quotidian fever as a temporary recovery.

General insanity being established, the burden of proof is thrown upon the party alleging a lucid interval, who must establish, beyond a mere cessation of the violent symptoms, a restoration of mind sufficient to enable the party soundly to judge of the act. Where a disease ultimately affecting the mind is insidious and slow in its development, and there is ground for suspicion that previous to the factum, apprehensions were entertained of the possible approach of mental derangement, there should be a careful scrutiny of an act performed shortly before an accession of undoubted symptoms, in order to see whether it was a rational and natural act comfortable with the views and wishes of the party when in a state of health.

Contracts or other legal instruments executed in alleged lucid intervals should therefore be closely scrutinized and looked upon with some degree of suspicion; but where capacity to contract is clearly shown to exist, the contract or other act must be sustained.

As to the degree of mental capacity requisite to make a valid will, Judge Redfield states that a lower degree of intellect is requisite to make a valid will than to make a valid contract; but in the former case something more is required than mere passive memory. There must be sufficient active memory to collect and retain the elements of the business to be performed for a sufficient time to perceive their obvious relations to each other. The testator must have a sound mind and disposing memory; in other words, he ought to be capable of making his will with an understanding of the nature of the business in which he is engaged, the elements of which the will is composed, and a recollection of the property of which he means to dispose, of the persons who are the objects of his bounty, and the manner in which it is to be distributed among them.

A morbid delusion is good in defeasance of a will

founded immediately in or upon such delusion. This was settled by the celebrated case of *Dew* vs. *Clarke*, decided by Sir John Nichol in the Prerogative Court of Canterbury in 1826. In this case Sir John Nichol stated his opinion that the true criterion of the absence or presence of insanity is the absence or presence of delusion. This, however, cannot at the present time be regarded as the law. If a delusion exists it is, of course, clearly evidence of insanity; but its absence is, as we have already seen, by no means proof or even evidence of a sane condition.

Deaf and Dumb Persons. — Deaf and dumb persons, although formerly in presumption of law idiots, are no longer so considered; and it may perhaps now be said that there is in the United States at least, as to them no presumption of a defective understanding. In order, however, to ensure protection and prevent fraud, proof would probably be required that such a person was capable of comprehending what he was about in executing any instrument.

Where a person presented as a witness is of unsound mind to such an extent as to be incapable of comprehending the nature and obligation of an oath, or the nature and relations of the subject matter about which he is to testify, it is very clear that he is not a competent witness. A person to some extent insane but not wholly devoid of reason, may, however, be permitted to testify if the court is satisfied that he possesses sufficient capacity to comprehend the facts, to understand the nature of an oath, and to communicate his testimony. Where such testimony has been received, evidence that the witness has been of unsound mind and memory is admissible to affect the credibility of his testimony; whether he has sufficient understanding must be decided by the court upon examination of the proposed witness himself and other witnesses who can speak as to the nature and extent of the insanity in question. Where the witness at the time of the trial is of sound mind and memory, but the evidence shows that his intellect was to a greater or less degree impaired at the time of the transaction as to which he testifies, he is a competent witness, but the question of his credibility is for the jury.

. See this whole subject considered at length in 16 "Western Jurist," 122.

Somnambulism. — As to the legal responsibility of persons accused of crime alleged to have been committed in the somnambulistic state, or in that state of mental confusion sometimes met with between sleeping and waking, the difficulty, as it seems to us, is principally one of proof. If it can be established to the satisfaction of the jury that the accused did not enjoy the free and rational exercise of his understanding, and was unconscious of his outward relations at the time of the commission of the alleged crime, there would seem to be no doubt that he is not criminally responsible.

Induced Hypnotism. — The subject of induced hypnotism is undergoing investigation both in France and in this country, and important questions as to criminal responsibility may arise therefrom. At present, however, in advance of any adjudicated cases upon the subject, its consideration here would be premature.

Drunkenness. — Voluntary drunkenness is no defence to a criminal charge. Where, however, a permanent condition of insanity has resulted from the habit of drunkenness, the same rule should be applied as in the case of insanity due to other causes. As to contracts, it is well settled that a contract entered into by a person when in such a state of intoxication as to deprive him of the exercise of his understanding is voidable, although the intoxication was voluntary and not procured through the circumvention of the other party.

A last will and testament made by a person while so intoxicated as not to understand the nature or effect thereof will not be allowed to stand.

CHAPTER XXII.

SOME RULES OF THE COMMON LAW RESPECTING THE
DISPOSITION OF HUMAN DEAD BODIES.

LEGISLATION UPON THE SUBJECT OF ANATOMY, AND A DRAUGHT
OF AN ACT TO PROMOTE THE SCIENCE OF ANATOMY,
MEDICINE, AND SURGERY.

So far as the writer has been able to discover, there
seems in all ages to have been, in the non-professional
mind at least, a peculiar aversion towards and horror of
dead bodies of human beings. The laws of Menu, enacted, according to Sir William Jones, from 880 to 1,280
years before Christ ("Sir William Jones's Works," pp.
79, 80), contain many provisions respecting uncleanness
and purification therefrom, by reason of the dead, and it
seems everywhere assumed that dead bodies are unclean.
Thus, among many other provisions, we find the following: "He who has touched a chandala, a woman in
her courses, an outcast for deadly sin, a new-born child,
a *corpse*, or one who has touched a corpse, is made pure
by bathing." — Laws of Menu, ch. 5, § 85.

"Should a Brahmin touch a human bone moist with
oil, he is purified by bathing; if it be not oily, by strok-
ing a cow, or by looking at the sun, having sprinkled
his mouth with water." — Ib. § 87.

The Koran likewise denounced as unclean the person
who touched a corpse, and the rules of Islamism still

forbid dissection. Likewise by the laws of the Franks, a person who dug a corpse out of the ground in order to strip it was banished from society, and no one suffered to relieve his wants till the relations of the deceased consented to his readmission. — 4 Black. Com., 235 ; Montesq. Spir. Laws, b. xxx. c. 19.

By the common law of England — and the rule of the American common law is believed to be the same — it is an indictable offence to take up a dead body, even for the purpose of dissection, as being "*contra bonos mores*, at the bare idea alone of which," say the court in *Rex* vs. *Lynn*, 2 Term, R. 733, " nature revolted."

It is said that a surgeon may retain the limbs he amputates from a patient, upon the ground that parts of the body when severed become dead, and at common law there is no property in a dead body.

While it is true that at the common law there can be no property in a corpse (see *Williams* vs. *Williams*, English High Court of Justice, Chancery Division, reported in 21 Am. Law Reg. [N. S.], August, 1882, p. 508, in the note to which the cases upon the subject are quite fully collected by the writer), and therefore stealing it is no felony, yet it is a very high misdemeanor at the common law, and at the present time " body-snatching," so-called, is made a statutory crime in probably every State of this Union.

In the case of Dr. Handyside, where trover was brought against him for the bodies of two children that grew together, Lord Chief Justice Willes held that the action would not lie, as no person had any property in corpses.

The case of *Williams* vs. *Williams* above cited, which referred to the subject of cremation, and in which it was held that a man cannot dispose of his body by will, that it is the executor's duty to bury it, and that meantime he has the right to possession of it, is an interesting case in this connection.

If, however, the coffin or any of the grave-clothes be stolen with the body, it was a felony at common law.

By the common law it is an offence against decency to take a person's dead body with intent to sell or dispose of it for gain or profit. Even to sell the dead body of a capital convict for the purposes of dissection, where dissection is no part of the sentence, is a misdemeanor and indictable at common law.

Even the refusal or neglect to bury dead bodies by those whose duty it is to perform the office appears also to have been considered a misdemeanor. So the prevention of the interment of a dead body has been considered indictable. To cast a dead body into a river has been held indictable at common law as an offence against common decency.

A gaoler has no right to detain the body of a person who has died in prison for any debts due to himself, and is indictable for so doing.

It seems that in a proper case the court, in the interest of justice, will order the disinterment and examination of a dead body, where there is good reason to believe that without such examination a fraud will be perpetrated, and where the defendant has exhausted all other legal methods of exposing it. One such case has recently come to the notice of the writer in the city of

Chicago, where the exhumation furnished conclusive evidence of an attempt to defraud an insurance company.

The dissection of human bodies being necessary for the advancement of anatomical and medical science, statutory enactments were made upon the subject in England at an early date, and also exist in many of the States in this country. The statutes of every State and Territory of this country have been examined with reference to this subject, and it was the writer's intention to have incorporated a digest of the same in this chapter, but it has been found that such a digest would consume more space than the merits of many of these statutes deserve, and it has been thought better to give the English legislation upon the subject, and at the close to submit a draught of an act containing the best features of them all.

The first English statute upon the subject of anatomy and dissection that we have been able to find is section 2 of chapter 42 of 32 Henry VIII., enacted in 1540, and entitled, "For Barbers and Surgeons." The second section of this act is as follows:—

"And further be it enacted by the authority aforesaid, that the said masters or governors of the mystery and commonalty of barbers and surgeons of London, and their successors yearly forever, after their said discretions, at their free liberty and pleasure, shall, and may have, and take without contradiction, four persons condemned, adjudged, and put to death for felony, by the due order of the King's laws of this realm, for anatomies, without further suit or labor to be made to the King's highness, his heirs or successors, (2) and to make incision of the same dead bodies, or otherwise to order the same after their said discretions at their pleasures, for their further and better knowledge, instruction, insight, learning, and experience in the said science or faculty of surgery." A similar grant was made by Elizabeth in 1565 to the College of Physicians.

By the statute 25 Geo. II., c. 37, 1752, entitled "An Act for better preventing the horrid crime of murder" (repealed by 9 Geo. IV., c. 31, § 1, 1828, for consolidating and amending the statutes in England relative to offences against the person; re-enacted in substance in sections 4 and 5 of the same statute, but repealed in 1838 by section 16 of 2 and 3 Wm. IV., ch. 75, which section 16 was itself repealed by 24 and 25 Vict., ch. 95, § 1), it was enacted, after providing (sect. 1) that murderers should be executed the next day but one after sentence passed, that (sect. 2) "the body of such murderer so convicted shall, if such conviction and execution shall be in the county of Middlesex, or within the city of London or the liberties thereof, be immediately conveyed by the sheriff or sheriffs, his or their deputy or deputies, and his or their officers, to the hall of the surgeons' company, or such other place as the said company shall appoint for this purpose, and be delivered to such person as the said company shall appoint for this purpose, and be delivered to such person as the said company shall depute or appoint, who shall give to the sheriff or sheriffs, his or their deputy or deputies, receipt for the same; and the body so delivered to the said company of surgeons shall be dissected and anatomized by the said surgeons or such person as they shall appoint for that purpose; and in case said conviction and execution shall happen to be in any other county or place in Great Britain,' then the judge or justice of assize, or other proper judge, shall award the sentence to be put in execution the next day but one after such conviction (except as before excepted), and the body of such murderer shall in like manner be delivered by the sheriff or his deputy, and his officers, to such surgeon as such judge or justice shall direct for the purpose aforesaid."

That the purpose of this act was less to advance the interests of science than to terrify wrong-doers is evident from section 3, which directs sentence to be pronounced immediately, stating the time of execution and the marks of infamy above specified, "in order to impress a just horror in the mind of the offender, and on the minds of such as shall be present, of the heinous crime of murder."

The statute above quoted has furnished a model, which, with more or less modifications, has been followed by several of the States of this Union, and among others, by the State of Illinois,—sect. 443 of whose criminal code provides that "the court may order, on the application of any respectable surgeon or surgeons, that the body of the convict shall after death be delivered to such surgeon or surgeons for dissection, unless the same be objected to by some relative of the convict." [1]

In 1832, the elaborate act of 2 and 3 Wm. IV., ch. 75, entitled, "An Act for Regulating Schools of Anatomy," was passed, the preamble of which is as follows:—

---

[1] See, however, the provisions of the act of 1835 ; ch. 91, R. S. Ill. 1885.

"Whereas, a knowledge of the causes and nature of sundry diseases which affect the body, and of the best methods of treating and curing such diseases, and of healing and repairing divers wounds and injuries to which the human frame is liable, cannot be acquired without the aid of anatomical examination; and whereas, the legal supply of human bodies for such anatomical examination is insufficient fully to provide the means of such knowledge; and whereas, in order further to supply human bodies for such purposes divers great and grievous crimes have been committed, and lately murder, for the single object of selling for such purposes the bodies of the persons so murdered; and whereas, therefore, it is highly expedient to give protection, under certain regulations, to the study and practice of anatomy, and to prevent, so far as may be, such great and grievous crimes and murder as aforesaid, be it enacted," etc.

The act then proceeds to provide (sect. 1) for licenses to practise anatomy; (sect. 2) for the appointment of inspectors of schools of anatomy; (sect. 3) the districts they shall superintend; (sect. 4) for returns by the inspectors of subjects removed for anatomical examination; (sect. 5) for the inspection of places where anatomy is practised; (sect. 6) for salaries of inspectors.

SECT. 7 makes it lawful for any executor or other person having lawful possession of the body of any deceased person, and not being an undertaker or other party entrusted with the body for the sole purpose of interment, to permit such dead body to undergo anatomical examination, unless, to the knowledge of such executor or other person such person during his life expressed a desire that his body should not undergo such examination, or unless the surviving husband or wife, or any known relative of the deceased, shall require the body to be interred without such examination.

SECT. 8 requires the party having lawful possession of the dead body of any deceased person who has, during his life, directed the anatomical examination of his body, to direct such examination to be made, unless the deceased person's surviving husband or wife, or nearest known relative, shall require the body to be interred without such examination.

SECT. 9 prohibits the removal for anatomical examination of the body of any person from the place where such person died, within a certain time, and without a certificate of the manner of death.

SECT. 10 makes it lawful for professors, surgeons, etc., being licensed as aforesaid, to receive bodies for anatomical examination under the provisions of the act.

SECT. 11 provides that the persons mentioned in the last section shall receive with the body a certificate as aforesaid, which shall be transmitted to the inspector of the district, with a return stating from whom received, date and place of death, sex, name, etc., and that the said certificate and particulars shall be by such license recorded in a book kept by him for that purpose, etc.

SECT. 12 requires notice to be given to the Secretary of State of places where anatomy is about to be practised.

SECT. 13 regulates the manner of removing bodies for examination, provides for their interment after examination, and for the transmission of a certificate of interment to the inspector of the district.

SECT. 14 provides that the persons licensed under the act shall not be liable to punishment for having in their possession or examining dead bodies according to the provisions of the act.

SECT. 15 provides that nothing in the act contained shall be construed to extend to or prohibit any post-mortem examination of any human dead body required or directed to be made by any competent legal authority.

SECT. 16 repeals so much of 9 Geo. IV., ch.. 31, as directs that the bodies of murderers may be dissected or hung in chains as ordered by the court; and enacts that such bodies shall be hung in chains or buried, as the court shall direct. (This sect. 16 was repealed by 24 and 25 Vict., c. 95, § 1).

SECT. 17 limits the time within which actions for anything done under the act shall be brought, and regulates the pleading therein.

SECT. 18 prescribes the punishment for offences against the act.

SECT. 19 defines the interpretation of certain words in the act.

SECT. 20 provides that the act shall go into effect August 1, 1832.

SECT. 21 provides that the act may be altered or amended during the current session of Parliament.

In 1871 (34 Vict., ch. 16, § 2) the Secretary of State and the chief Secretary of Ireland were empowered from time to time to vary the period limited by sect. 13 of 2 and 3 Wm. IV., ch. 75, for transmission of certificates of interment to district inspectors.

This appears to be the latest legislation upon the subject in England.

In this country there is quite a diversity among the statutes upon this subject, and in many of the States, especially the Southern States, there is no legislation whatever upon the subject, other than statutes prohibiting the robbing of graves, etc. Many statutes provide that "it shall be lawful" for the designated officers to deliver up the dead bodies, etc., or that they "may" deliver, etc., without containing words making it the imperative duty of such officers to deliver, etc. It is very possible, and in some cases probable, that the word "may" in such statutes should be construed to mean "shall" or "must." "The words 'may' or 'shall,' when used in a statute, may be read interchangeably, as will best express the legislative intention. The rule is that 'the word *may* means *must* or *shall* only in cases where public interests and rights are concerned, and the public or third persons have a claim *de jure* that the power shall be exercised.'" Where a statute directs the doing of a thing for the sake of justice or the public good, the word "may" is the same as the word "shall." Thus, where a statute says that a sheriff *may* take bail, it has been construed to mean that he shall do so.

On the other hand, if any right to any one depends upon giving to the word "shall" an imperative construction, the presumption is that the word was used in reference to such right or benefit; but where no right or benefit to any one depends upon the imperative use of the word, it may be held to be directory merely.

In the draught of a statute upon this subject, therefore, it is better, in order to avoid doubt, to use such words as express unequivocally the intention of the legislature to impose upon the officials mentioned in the act a duty to deliver, etc.

Many statutes contain a provision authorizing the delivery of dead bodies to county medical associations, or to some reputable physician in cases where there is no medical school in the county. Such a provision, as well as others that might be mentioned, are doubtless desirable in some cases, but they have not been included in this draught. If thought desirable, it will be easy to incorporate such provisions. It is believed that all the really good features of the various acts upon this subject, which are to be found upon the statute-books of the various States of this Union, will be found in this bill, and a number of other provisions which seemed necessary have also been incorporated in it. It is inserted in this place simply as a precedent, in the hope that it may be of assistance to some one called upon to draught such an act.

An Act to promote the science of anatomy, medicine, and surgery.

Be it enacted, etc. : —

SECTION 1. The right to dissect the dead body of a human being or any part thereof shall exist in the following cases :

(1) In cases authorized by positive enactment of the General Assembly of this State, in this and other statutes.

(2) Whenever a coroner is authorized by law to hold an inquest upon the body, so far as such coroner authorizes dissection for the purposes of the inquest and no farther.

(3) Whenever and so far as the husband, wife, or next of kin of the deceased, in case the deceased leaves no surviving husband or wife, may authorize dissection for the purpose of ascertaining the cause of death, and no farther.

(4) Whenever a person has, during his lifetime directed that his dead body, or any part thereof, may or shall be dissected, or where any person has directed or given permission that any part of his body which has become separated therefrom during his lifetime be dissected, such dissection shall be lawful to the extent authorized by such person, and no farther.

SECT. 2. Every person who makes, or procures to be made, any dissection of the dead body of a human being, or of any part thereof, contrary to the provisions of this act, shall be deemed guilty of a misdemeanor, and upon conviction thereof shall be fined not less than $25 nor more than $100, or be imprisoned in the county jail not less than one month nor more than three months, or shall be punished by both said fine and imprisonment, at the discretion of the court.

SECT. 3. Every superintendent of a penitentiary, State, city, or county hospital for the insane, warden of poor-house, coroner, sheriff, city undertaker, and every other public officer by whatsoever name designated, lawfully having charge of the body of any deceased person required to be buried at the public expense, shall immediately, by telegraph when practicable, otherwise by letter, notify the nearest known relative of such deceased person, if he knows or can with reasonable diligence ascertain the same, or, if no relative can be found, then some personal friend of such deceased person, if any such is known to exist, of the death of such person, and shall deliver the body of such deceased person to any such relative or personal friend, who is known or shall prove himself to be such to the officer or authorities having charge of the body, and who shall claim the same for interment within a reasonable time after such notice, not exceeding forty-eight hours after the death of such person; but if no such relative or friend shall claim the body within forty-eight hours after death, it shall be the duty of such officer or authorities having charge of such body forthwith to deliver the same to the officer or duly authorized agent of any respectable medical college, of whatever school, regularly chartered by this State, making application therefor, and which has given the bond hereinafter referred to, and otherwise complied with the requirements of this act necessary to obtain its benefits; and it shall be lawful for such officer or agent to receive the said dead body, and for the said medical college, through its professors, officers, and students, to use the same as they may deem most for the advancement of anatomical medical, and surgical science, but for such purposes only, and in this State only.

Provided, That if any body so delivered shall be subsequently claimed for interment by any such relative or friend of the deceased, it or the remains thereof shall be forthwith surrendered up for that purpose.

Provided, also, That the remains of no deceased person who is known during his last sickness to have expressed a desire to be buried, the remains of no traveller dying suddenly, the remains of no person detained on any civil process, or as a witness, shall be so delivered, but shall be decently buried in the usual manner.

Provided, also, That the bodies so delivered shall be distributed among the several respectable, regularly chartered medical colleges of this State, of whatsoever school of medicine, according to the number of students regularly matriculatéd during the winter session of the college year next preceding the time of application and distribution; provided, however, that no medical college shall receive more than three (3) bodies, till every other medical college which has filed the bond hereinafter required, otherwise complied with the requirements of this act, and made application therefor, shall have received at least one (1); and no application shall be made, or if made, shall be allowed, till the college so applying is ready to receive and remove the body or bodies applied for, and has given the bond hereinafter required and otherwise complied with the requirements of this act. Any violation of any provision of this section shall be deemed a misdemeanor, and shall be punished by a fine of not less than $50 nor more than $200, or by imprisonment in the county jail not less than three nor more than six months, or by both said fine and imprisonment at the discretion of the court.

SECT. 4.   In order to facilitate the equitable distribution of said bodies, it shall be the duty of every medical college so applying to file with each and every application for a body or bodies a statement of the number of students matriculated as aforesaid, and the number of bodies received by it from all sources up to the date of such application under the provisions of this act during the current college year, which for the purposes of this act

shall be deemed to commence upon the day when this act
goes into effect, and end on the 31st day of August next
ensuing, and every year thereafter shall commence with
the 1st day of September and end on the 31st day of
August of each and every year; and no application shall
be allowed which is not accompanied by such statements.
In making the distribution and delivery of dead bodies
above provided for, each and any officer or person concerned in the same shall be governed by the statements
thus filed and by the principles stated in this and the
preceding sections. In order to equalize any irregularity of distribution resulting therefrom, it shall be the
duty of each and every medical college claiming the
benefit of this act, through its clerk or other recording
officer, on or before the 5th day of September, in each and
every year, to file with the Secretary of the State Board
of Health a statement, in writing and under oath, of the
number of students regularly matriculated during the
winter session of the college year ending on the 31st
day of August next preceding, and the entire number of
bodies received from all sources under this act during
the same period; and it shall be the duty of the said
Secretary of the State Board of Health forthwith to
equalize the distribution of bodies appearing from such
statements, according to the principles of this act, and
to certify the results of such equalization and transmit
a copy thereof under his hand and seal as soon as may
be to each college which has filed the statements above
required, showing (1) the entire number of bodies each
college was entitled to receive under this act during the
year next preceding; (2) the number of bodies each college has in fact received; and (3) the number of bodies

to be delivered by each college respectively which has received an excess over the number it was entitled to receive to each college respectively which has received a less number than it was entitled to receive; and thereupon it shall be the duty of each college respectively thus appearing to be in excess to deliver, within a reasonable time after the reception of such copy of said certificate, to each college respectively which has received less than the number it was entitled to receive, so many bodies as said certificate shall specify as being necessary to equalize .the distribution according to the principles hereinbefore laid down.   Every medical college neglecting or refusing to file with the Secretary of the State Board of Health the statement above required, or neglecting or refusing to comply with the terms of equalization above provided for, as shown by the said certificate of the Secretary of the State Board of Health, shall thereby forfeit during the time it shall so neglect or refuse such compliance all rights and benefits otherwise accruing to it under this act; and a subsequent compliance with such requirements shall not entitle it to the benefits which might have otherwise accrued to it during the period it was so in default.

SECT. 5.   Every medical college claiming the benefit of this act shall, before it shall be entitled to receive any dead body as aforesaid, execute and file with the clerk of the county in which such college is situated, who shall give a receipt therefor, a bond to the People of the State of . ... in the penal sum of $1,000, with a surety or sureties to be approved by said county clerk, conditioned that each and every body received under the provisions of this act shall be used only for the advancement of anatomical, medical, and surgical science, in this State only, and in such a manner as not to shock the public sensibilities; that said college will cause to be kept the record required by this act; and that the remains of every such body, after use as aforesaid, shall be decently buried in some public burial-ground in this State; which bond shall be renewed on the first day of September of each and every ensuing year.

. SECT. 6. The person receiving any dead body under this act shall, in the name and in behalf of the medical college for which he receives it, sign and deliver to the officer or person from whom the same is received, and whose duty it shall be to demand and obtain such receipt, a receipt therefor, stating, if known, the name, age and sex of every such person, and the place, date, and cause of death, if known; which receipt shall be preserved and recorded in a book to be kept for that purpose in the institution, association, or office from which such body shall be delivered, and a copy thereof immediately transmitted to the Secretary of the State Board of Health, whose duty it shall be to file the same in his office. And every medical college receiving any dead body under this act shall by its demonstrator of anatomy or other analogous officer, in a suitable book to be kept for that purpose, make a legible record of the time when, the name and official station of the person from whom, and the place where, such body was received, and whether or not such body when so received was inclosed in any box, cask, or other receptacle, and if so inclosed, such record shall contain a description of such box, cask, or receptacle sufficient to identify the same, together with the shipping mark, or directions, if any, on the same. Such record shall contain a description of such body or remains, including the name, if known, sex, length and weight of the body, and the probable age of the deceased at the time of death, color of hair and beard, if any, condition of teeth, and any and all wounds, marks, or scars, if any, on such body by which the same might be identified; and whether or not such body when so received was mutilated so as to preyent identification of the same. And such record shall be preserved by such medical college through its demonstrator of anatomy or other analogous officer, whose duty it shall be to exhibit the same on demand, as also any and all such dead bodies then in his charge, for the inspection of any sheriff or deputy sheriff of this State. Any violation of any provision of this section shall be deemed a misdemeanor, and shall be punished by a fine not of less than $50 nor more than $200, or by imprisonment in the county jail not less than three months nor more than six months, or by both said fine and imprisonment at the discretion of the court.

SECT. 7. Any person who shall buy or offer to buy, sell or offer to sell, the dead body of any human being or procure the same to be done by another; or any person who shall offer, pay, demand, or receive any money or any valuable consideration whatever, or procure the same to be done by another, in consideration of the delivery for anatomical, medical, or surgical purposes, of the dead body of any human being, or who shall transport the dead body of any human being beyond the limits of this State for anatomical, medical or surgical purposes, or who shall procure the same to be done by another, shall be deemed guilty of a felony, and upon conviction thereof shall be punished by imprisonment in the State's Prison not less than one year nor more than five years.

INDEX.

INDEX.

University Press: John Wilson & Son, Cambridge.

www.ingramcontent.com/pod-product-compliance
Lightning Source LLC
Chambersburg PA
CBHW021348210326
41599CB00011B/794